# Advances in Industrial Control

Springer
*London*
*Berlin*
*Heidelberg*
*New York*
*Barcelona*
*Hong Kong*
*Milan*
*Paris*
*Santa Clara*
*Singapore*
*Tokyo*

*Other titles published in this Series:*

*Neural Network Engineering in Dynamic Control Systems*
Edited by Kenneth Hunt, George Irwin and Kevin Warwick

*Neuro-Control and its Applications*
Sigeru Omatu, Marzuki Khalid and Rubiyah Yusof

*Energy Efficient Train Control*
P.G. Howlett and P.J. Pudney

*Hierarchical Power Systems Control: Its Value in a Changing Industry*
Marija D. Ilic and Shell Liu

*System Identification and Robust Control*
Steen Tøffner-Clausen

*Genetic Algorithms for Control and Signal Processing*
K.F. Man, K.S. Tang, S. Kwong and W.A. Halang

*Advanced Control of Solar Plants*
E.F. Camacho, M. Berenguel and F.R. Rubio

*Control of Modern Integrated Power Systems*
E. Mariani and S.S. Murthy

*Advanced Load Dispatch for Power Systems: Principles, Practices and Economies*
E. Mariani and S.S. Murthy

*Supervision and Control for Industrial Processes*
Björn Sohlberg

*Modelling and Simulation of Human Behaviour in System Control*
Pietro Carlo Cacciabue

*Modelling and Identification in Robotics*
Krzysztof Kozlowski

*Spacecraft Navigation and Guidance*
Maxwell Noton

*Robust Estimation and Failure Detection*
Rami Mangoubi

*Adaptive Internal Model Control*
Aniruddha Datta

*Price-Based Commitment Decisions in the Electricity Market*
Eric Allen and Marija Ilić

*Compressor Surge and Rotating Stall*
Jan Tommy Gravdahl and Olav Egeland

*Radiotherapy Treatment Planning*
Oliver Haas

Pushkin Kachroo and Kaan Özbay

# Feedback Control Theory for Dynamic Traffic Assignment

With 100 Figures

Springer

Pushkin Kachroo, PhD
Bradley Department of Electrical and Computer Engineering,
Virginia Polytechnic Institute and State University,
Blacksburg, Virginia 24061-0111, USA

Kaan Özbay, PhD
Department of Civil and Environmental Engineering, Rutgers University,
623 Browser Road, Piscataway, New Jersey 08854-8014, USA

ISBN-13: 978-1-4471-1209-9 Springer-Verlag London Berlin Heidelberg

British Library Cataloguing in Publication Data
Kachroo, Pushkin
Feedback control theory for dynamic traffic assignment.
(Advances in industrial control)
1. Traffic assignment 2. Control theory
I. Title II. Özbay, Kaan
388'.041
ISBN-13: 978-1-4471-1209-9

Library of Congress Cataloging-in-Publication Data
Kachroo, Pushkin.
Feedback control theory for dynamic traffic assignment. / Pushkin Kachroo and Kaan Özbay.
p. cm. — (Advances in industrial control)
ISBN-13: 978-1-4471-1209-9 e-ISBN-13: 978-1-4471-0815-3
DOI:10.1007/978-1-4471-0815-3
1. Electronic traffic controls. 2. Adaptive control systems.
I. Özbay, Kaan. II. Title. III. Series.
TE228.K33 1999 98-41884
388.3'122—dc21 CIP


Softcover reprint of the hardcover 1st edition 1999



Typesetting: Camera ready by authors

69/3830-543210 Printed on acid-free paper

# Advances in Industrial Control

Professor Dr -Ing M. Thoma
Westermannweg 7
D-30419 Hannover
Germany

Professor H. Kimura
Department of Mathematical Engineering and Information Physics
Faculty of Engineering
The University of Tokyo
7-3-1 Hongo
Bunkyo Ku
Tokyo 113
Japan

Professor A.J. Laub
College of Engineering - Dean's Office
University of California
One Shields Avenue
Davis
California 95616-5294
United States of America

Professor J.B. Moore
Department of Systems Engineering
The Australian National University
Research School of Physical Sciences
GPO Box 4
Canberra
ACT 2601
Australia

Dr M.K. Masten
Texas Instruments
2309 Northcrest
Plano
TX 75075
United States of America

Professor Ton Backx
AspenTech Europe B.V.
De Waal 32
NL-5684 PH Best
The Netherlands

To

my Dad (Dr. P. L. Kachroo)
my Mom (Sadhna Kachroo)
my brother (Dhananjaya Kachroo)
my wife (Anjala S. Krishen),
my children (Axenya Kachen and Sheen Kachen), also
Myshkin Kachroo, Parmesh and Mahesh

Dr. Kumar Krishen
Vijay Krishen
Sweetie Krishen
Lovely
Sinan
Kaan Ozbay

Pushkin Kachroo

To

my Dad (Dr. [illegible] Kachroo)
my Mom (Sadhna Kachroo)
my brother ([illegible] Kachroo)
my wife (Anjali S. Kachroo)
[illegible]
[illegible]

[illegible]
Vijay Krishna
[illegible]
[illegible]

## SERIES EDITORS' FOREWORD

The series *Advances in Industrial Control* aims to report and encourage technology transfer in control engineering. The rapid development of control technology impacts all areas of the control discipline. New theory, new controllers, actuators, sensors, new industrial processes, computer methods, new applications, new philosophies,...., new challenges. Much of this development work resides in industrial reports, feasibility study papers and the reports of advanced collaborative projects. The series offers an opportunity for researchers to present an extended exposition of such new work in all aspects of industrial control for wider and rapid dissemination.

Micro-technology and modern communications technology are revolutionising many aspects of our daily lives and so it is not surprising that it is impacting societal transportation systems whether our highways, airways, seaways or railways. The Advances in Industrial Control series reported on these developments for long haul railway systems in a monograph by Howlett and Pudney (ISBN 3-540-19990-X, 1995). Now it is the turn of transportation in a contribution from Pushkin Kachroo and Kaan Özbay.

The authors viewpoint is that this new set of transportation problems are control problems and that control engineers should be highly active in this field. Their volume covers all the aspects of modelling, problem formulation, and applies various control methodologies to solve the control problems formulated.

The monograph is likely to have wide appeal to control systems theorists, control engineers, mathematician and transportation engineers. An unusual feature of the monograph is the provision of Questions and Problems sections so the volume has a pedagogical value too!

M.J. Grimble and M.A. Johnson
Industrial Control Centre
Glasgow, Scotland, UK

# PREFACE

Traditionally, traffic control problems have been studied from the planning perspective. However, due to the advancements in micro-technology in terms of availability of various sensors, actuators, processors, and communication hardware, it has become possible to perform real-time control of traffic. It is important that we design and analyze these closed-loop systems from the perspective of feedback control theory, since these systems are in fact feedback control systems.

This book can be read by researchers and students having different backgrounds. We have attempted to provide as much basic information as possible. However, if we provided all the information to make this book completely self-sufficient, the size of the book would at least double. The references given at the end of each chapter should provide ample material as background material and also as topics of further research. This book should be of interest to people from traditional civil engineering, from control engineering (electrical, computer and mechanical engineering), and from applied mathematics. Readers with a transportation background might read the book following a different sequence than other readers. For example, a reader with a transportation background might skim through chapter 2, but readers lacking that background would read that chapter more carefully.

We were inspired by the initial work of Markos Papageorgiou in using feedback control theory for traffic control problems. The Center for Transportation Research (CTR) at Virginia Tech. provided us with the ideal opportunity to study the design of feedback controllers for various traffic control problems, which include traffic routing/assignment, ramp metering and signalized intersection. Professor Antoine G. Hobeika was the director of CTR when we started the research. After that, Ray D. Pethtel supported our research a great deal as the interim director, following which Professor Thomas A. Dingus continued the support as he took over the directorship of CTR. The support for the basic research out of which this book evolved, came through many projects at CTR mostly funded by the Research Center of Excellence (RCE) from Federal Highway Administration (FHWA). Virginia Department of Transportation (VDOT) has provided research support and the opportunity to study its working traffic control systems, which proved very useful. In 1997, Professor Michael W. Van Aerde, the creator of INTEGRATION traffic simulation software, Dr. Hesham A. Rakha, and Dr. Wei H. Lin joined CTR. Their immense knowledge and experience has been beneficial for the first author of this book. From FHWA, we would like to acknowledge the encouragement which was provided by Dave G. Gibson, and from VDOT, Mr. Jim Robinson, and Mr. David Gehr. Our collaboration with University of Virginia gave us some important perspective on traffic operation, and Dr. Brian Smith and Dr. Gary Allen provided that.

The first author has learnt a great deal from his association with his friends and colleagues Professor Joseph A. Ball and Professor Martin V. Day from the department of mathematics at Virginia Tech. He hopes that one day he too can acquire the same analytic skills possessed by the two. Dr. Mehdi Ahmadian has been a great friend who has helped me through different times and provided encouragement.

Most importantly, for the first author, the support given to him by his mentors at University of California at Berkeley can not be overemphasized. His advisor Professor Masayoshi Tomizuka has provided a foundation for conducting research, for which he is indebted to him. Professor Pravin Varaiya with his tremendous expertise in traffic systems, communication systems and power systems, to name just a few, has become a role model for the first author. I have also been fortunate to have had the guidance of Professor S. L. Dhingra at I.I.T. Bombay during my B.Tech.

The authors would like to acknowledge the help that Trina F. Murphy gave in reviewing and finalizing this book. We strongly believe that she is one of the best in her field. In addition, we would like to thank Michael J. Anderson for his assistance during the final stages of document production.

This book is a compilation and enhancement of the work presented by the authors in the following papers:

1. Pushkin Kachroo, Kaan Özbay, Sungkwon Kang, and John A. Burns, "System Dynamics and Feedback Control Formulations for Real Time Dynamic Traffic Routing with an Application Example" Mathl. Comput. Modelling Vol. 27, No. 99-11, pp. 27-49, 1998.
2. Pushkin Kachroo and Kaan Özbay, " Solution to the User Equilibrium Dynamic Traffic Routing Problem using Feedback Linearization ", Transportation Research: Part B, Vol. 32, No. 5, pp. 343-360, 1998.
3. Pushkin Kachroo and Kaan Özbay, "Real Time Dynamic Traffic Routing Based on Fuzzy Feedback Control Methodology", Transportation Research Record 1556, 1996.
4. Pushkin Kachroo, and Masayoshi Tomizuka, "Chattering Reduction and Error Convergence in the Sliding Mode Control of a Class of Nonlinear Systems", IEEE Transactions on Automatic Control, vol. 41, no. 7, July 1996.
5. Pushkin Kachroo, and Masayoshi Tomizuka, "Integral Action for Chattering Reduction and Error Convergence in Sliding Mode Control", American Control Conference, Chicago, 1992.
6. Pushkin Kachroo, Kaan Özbay, and Arvind Narayanan "Investigating the Use of Kalman Filtering Approaches for Origin Destination Trip Table

Estimation", Proceedings of IEEE Southeastcon '97, Blacksburg, VA, April 12-14, 1997.

7. Pushkin Kachroo, and Kaan Özbay "Feedback Control Solutions to Network Level User-Equilibrium Real-Time Dynamic Traffic Assignment Problems", Proceedings of IEEE Southeastcon '97, Blacksburg, VA, April 12-14, 1997.
8. Pushkin Kachroo and Kaan Özbay, "Sliding Mode for User Equilibrium Dynamic Traffic Routing Control", Proceedings of IEEE Conference on Intelligent Transportation Systems ITSC'97, Boston, 1997.
9. Pushkin Kachroo and Kumar Krishen, "Feedback Control Design for Intelligent Transportation Systems," Proceedings of Third World Conference on Integrated Design & Process Technology, July 6-9, 1998, Berlin.

Some material is reprinted with permission of Virginia Department of Transportation , and some material, especially in chapter 3, is derived from paper 1 (above) Copyright (1998) and reprinted with permission of Elsevier.

**Pushkin Kachroo**
Bradley Department of Electrical and Computer Engineering,
Virginia Polytechnic Institute & State University,
Blacksburg, VA 24061
Also:
Center for Transportation Research,
Virginia Polytechnic Institute & State University,
Blacksburg, VA 24061

**Kaan Ozbay,**
Department of Civil and Environmental Engineering
Rutgers University
Piscataway, New Jersey, 08854

# CONTENTS

# CHAPTER 1
# INTRODUCTION

## Objectives

To provide the motivation behind using feedback control for dynamic traffic diversion and dynamic traffic assignment
Provide some preliminary ideas about feedback control methodology
Describe how traffic control systems are set up for traffic management purposes

## 1. Dynamic Traffic Routing

Dynamic traffic routing (DTR) refers to the process of diverting traffic at a junction dynamically. Static diversion would be the case when the amount of traffic to be diverted has been pre-calculated and does not change with time. Being dynamic implies that the values change with time as the traffic conditions also change. Figure 1.1 shows a sample site where dynamic traffic routing would be highly beneficial. This is a traffic site where most of the traffic travels from point A to point B in the morning rush hours and in the reverse direction during the evening hours. There are two highways, which are connected at point A. Vehicles that want to go to point B can take either of the two highways, and normally they take a comparable amount of time. If there is congestion in one of the routes, then the travel time on that route will increase. Hence, more traffic should be diverted onto the other route. In general, if the travel time is the same on both routes, we can claim that this traffic system is working well. Therefore, when we develop traffic controllers in this book, we will keep equal travel time as one of the objectives of the control. This objective is called "user equilibrium" since users try to or would like to emulate that kind of route choice behavior to obtain maximum benefit. Another objective would be to obtain "system optimal," which means that for the total traffic network, the overall time created by using the specific choice of traffic diversions at all traffic nodes is optimal (implying that it is less than the total travel time created by any other choice of diversion strategy).

Traffic assignment has many applications. It is used in the following three ways in traffic applications:

(1) **Transportation Planning**: It is part of transportation planning process which includes travel demand analysis, travel forecasting, trip generation, trip distribution, mode choice and traffic assignment. Transportation planning process might be used to see the impact of a proposed new road, or new changes to an area, etc.
(2) **Simulations**: Traffic simulations are used to evaluate various traffic control measures and are also be used to analyze impact of transportation planning. Traffic simulations usually depend on the given origin-destination (O-D) travel demands and based on those traffic assignment is performed on various routes of all O-D pairs.

(3) **Real-time Traffic Control**: Real-time traffic control refers to the actual control of real traffic in real-time in order to affect the traffic behavior and performance.

Feedback control is extremely useful and important in real-time traffic control. However, it can be used effectively in transportation planning and simulation assignment problems also.

There are two types of traffic assignment techniques: static and dynamic and there are discussed further below:

(1) **Static Traffic assignment**: Traditionally for transportation planning purpose static traffic assignment technique has been used. In this technique assignment is done without regard to time as a variable. There are essentially three basic techniques used for this: (a) Diversion curves which give a relationship between the percent traffic split between alternate routes and the ration of travel time on the routes, (b) minimum time path (all or nothing) assignment, in which all the traffic is assigned to the minimum time path between the origin and the destination node, and (c) minimum time path with capacity constraint, in which travel times are adjusted on links after minimum time path assignment, and the solution is obtained after multiple iterations.

(2) **Dynamic Traffic Assignment (DTA)**: Dynamic traffic assignment is traffic assignment where time is a variable which is used in the assignment and modeling. For instance, in simulations, the O-D demands are time-dependent. In real-time traffic control, traffic has to be assigned dynamically to achieve some traffic performance.

**Objectives**

In summary, traffic controllers are designed to satisfy one of the following objectives:

User Equilibrium: Travel time on alternate routes should be the same.

System Optimal: The total travel time on the system should be the minimum.

**Figure 1.1:** *A Sample Site for Diversion (Courtesy of VDOT)*

## Control Algorithm Design

Now the question arises regarding how we can achieve the right amount of diversion in order to equate the travel times on the two alternate routes. One issue is the calculations of the right split factor (that is, the percent of traffic flow entering each alternate route compared to the total traffic flow coming to the node). For a real-time traffic responsive system, the split factors should be functions of instantaneous traffic conditions (such as traffic densities, flow, or traffic speed at various locations on the routes). For instance, if, in general, traffic density is greater on one route at some time, then instantaneously we could try to change the split factor so that more traffic goes to other alternate routes. The development of algorithms that calculate the split factor values that are functions of the traffic variables is the main concern of this book and is presented in various following chapters.

## Sensing

In order to obtain the values of the split factors in real-time as functions of traffic variables, the traffic variables need to be measured. Many types of traffic sensors are used for that purpose. Traffic sensors can use many technologies such as video cameras, loop detectors, piezoelectric sensor, fiberoptics, etc. Figure 1.2 shows the internet site of the Hampton Roads area maintained by the Virginia Department of Transportation (VDOT), which gives real-time data obtained by the traffic sensors in that area. The same data is also used for traffic control. Figure 1.3 shows a traffic camera used in the Transguide System of the Texas Department of Transportation (TxDOT).

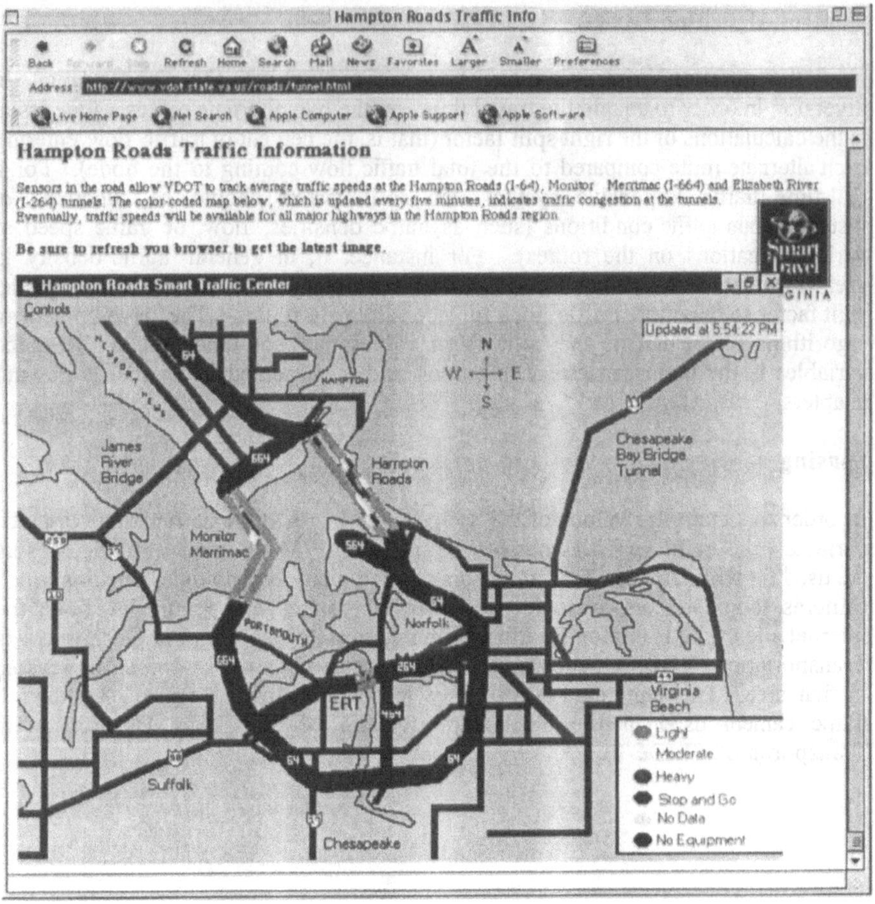

**Figure 1.2:** *Virginia Department of Transportation Internet Site for Hampton Roads Area (Courtesy of VDOT)*

**Figure 1.3:** *Traffic Camera Used in the Transguide System of the Texas Department of Transportation (TxDOT) in San Antonio (Courtesy of TxDOT)*

**Actuation**

After the control algorithm calculates the split factor desired at each traffic node, this split factor value is implemented. There are various actuation methods, which can be applied for this purpose. Some of these are: Variable Message Signs (VMS; see Figure 1.4), Highway Advisory Radio (HAR), and In-vehicle Information.

**Figure 1.4:** *Variable Message Sign in the Transguide System (Courtesy of TxDOT*

**Automatic Control vs. Human-in-the-loop Control**

After the data is acquired by sensors, it is processed by computers and can then be displayed to operators in a traffic center. The operators can then decide on traffic control measures such as what signs to display on VMS or what signal timing control strategies to use. In automatic mode, some of the functions of the human operators could be bypassed. For instance, if research would give clear indication as to how VMS signs affect the split factors at traffic nodes, then the feedback control algorithm would calculate the split factors automatically based on the measured data, and then drive the VMS signs. The split factor data can be used by operators to determine the control measures based on the split factor values calculated by the algorithms.

**Overall System**

An example of an overall traffic control system is the Transguide System being deployed in San Antonio by TxDOT. The details of this system are presented at the end of this chapter to give the reader an idea of how traffic control systems are practically implemented.

**Traffic Analysis Notation**

For analysis of traffic networks, we show highways by lines (called links); where two or more lines connect, we show that point by a dot and call it a traffic node (as shown in Figure 1.5). Links can be two-way (allowing traffic in both directions) or one-way. Nodes can be origin nodes (indicating that traffic is originating from them), destination nodes, or intermediate nodes. More details on these will be given in subsequent chapters.

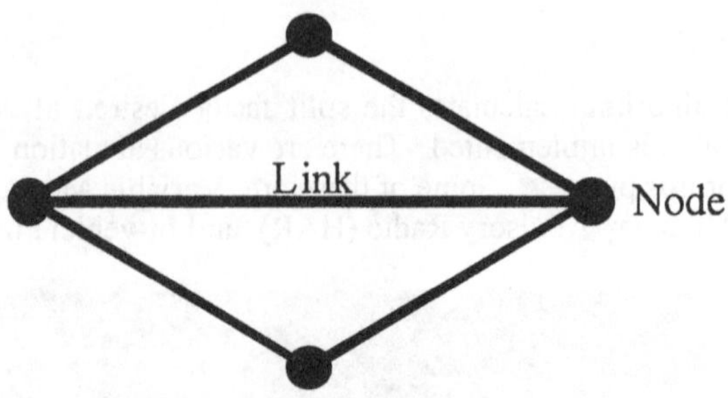


**Figure 1.5:** *Traffic Network Representation*

## 2. Motivation

Dynamic Traffic Routing and Assignment (DTR/DTA) has been one of the important research topics in transportation engineering. Although there have been several recent attempts to develop real-time Dynamic Traffic Assignment/Routing (DTA/DTR) algorithms that will perform on-line traffic control, the majority of the transportation research has focused on off-line planning problems. With the advent of Intelligent Transportation Systems (ITS), the need for dynamic models capable of working in real-time has become clearer.

The traditional optimization-based approaches attempt to solve the DTR/DTA problem by optimizing the objective functions for the nominal model over the "planning horizon." For real-time traffic flow control, where on-line sensor information and actuation methods are available, this technique is not very well suited. On the other hand, there exist real-time feedback control approaches that are specifically designed for such systems. The major motivation of this paper is to discuss possible ways of modeling the DTR problem in order to facilitate the design of feedback control laws. The drawback of the linear feedback control techniques that have been tried is that the system should remain in the linear region at all times for the controller to be valid. Since the system is nonlinear, time varying, and contains uncertainties, feedback control laws that handle such systems should be developed.

In this book, we will develop a design methodology for feedback control laws for DTR. Although the original system is infinite dimensional, the control input is finite dimensional. The feedback process is illustrated in Figure 1.6. There are essentially three ways the system modeling could be used to design feedback controllers. These are:

(1) Distributed Parameter setting, represented by Partial Differential Equations (PDE).
(2) Continuous Time Lumped Parameter setting, represented by continuous time Ordinary Differential Equations (ODE).
(3) Discrete Time Lumped Parameter setting, represented by continuous time ODEs.

We will present feedback control formulations to reach user-equilibrium in these three settings.

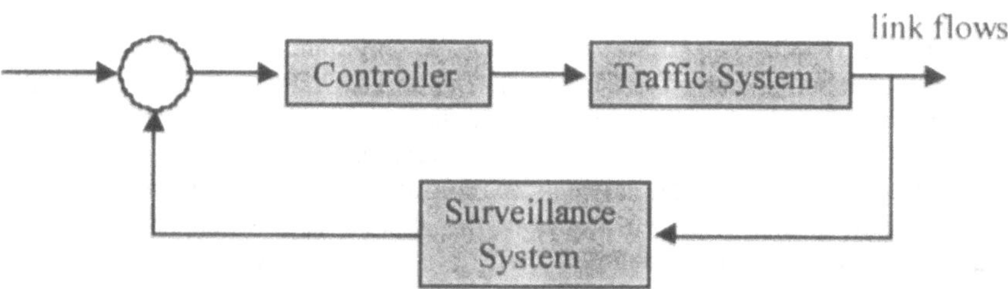


**Figure 1.6:** *Block Diagram for DTR Feedback Control*

There are various advantages and disadvantages in designing the feedback controller using any one of the three kinds of models. The original PDE model is derived from the hydrodynamic analogy presented by Lighthill and Whitham [1]. It is difficult, however, to design a feedback controller directly for a distributed parameter system, and it is an area of active research. By space discretizing the model, we can design a feedback controller in the continuous time domain which can be easier to design. This model obviously will have discretization errors, which could be reduced, however, by designing a robust controller that would attempt to eliminate these errors. Finally, it is natural to design a controller using a discrete time ODE model of the system for discrete implementation of the control. Again, this model would have more discretization errors, and the controller would have to minimize the effect of those.

The question of moving the traffic bottleneck from one route to another as a result of traffic diversion is an important one. The risk of creating a bottleneck on an alternate route to which the traffic is diverted can be essentially due to the overreaction of drivers who divert in large numbers to the alternate route(s) in response to the routing advisory.

If an open loop system that uses off-line optimization and/or simulation models is employed to determine the traffic control measures, the control system will not be able to capture this overreaction and respond to it effectively. This in turn will result in moving the bottleneck from one route to another. However, a closed loop system similar to the one proposed in this paper will capture any user overreaction and revise its routing control commands in order to prevent the occurrence of a bottleneck on the alternate routes. However, the effectiveness of the prevention of the creation of new bottlenecks due to the user overreaction will depend on the robustness of the controller.

## 3. Literature Review

Real-time diversion and routing of traffic is one of the efficient ways of relieving non-recurrent traffic congestion. Several models have been developed for determining diversion routes and diverting the traffic onto these routes. Expert systems, feedback control, and mathematical programming models are among the approaches that have been used for developing real-time diversion and routing strategies.

A multivariable feedback regulator with integral parts and a simple bang-bang controller have been developed and tested on a particular traffic network model [2]. This approach is attractive, but is difficult to design for nonlinear systems. One approach is to linearize the plant, but then the results will be valid only in the linear region. Another method suggests the application of Linear Quadratic (LQ) optimal control on the linearized model of the system [3], which has the same difficulty. In one approach, feedback control is realized by solving nonlinear optimization using gradient search over a sufficiently long future time horizon at each control interval [4, 5]. This method obviously involves more computation and also might generate difficulty with the proof of performance characteristics.

Gupta et al. [6] developed a framework for freeway incident management. In this framework, an expert system first fixes the appropriate strategy to manage the incident; then if diversion is felt to be necessary, control is passed onto the diversion algorithm. Diversion routes are not dynamically generated, but a predetermined set of routes are used for analysis.

Gartner and Reiss [7] have developed a traffic control system design for congestion control in freeway corridors. This system, called the Integrated Motorist Information System (IMIS), is for freeway and arterial traffic management on 128 miles of heavily traveled highway in New York. The traffic control system serves to rapidly detect congestion and implements controls to minimize traffic disruptions. The possible controls include diversion, ramp metering, and signal retiming. The most important element of the diversion framework developed is the diversion control module. This control module is structured hierarchically as corridor-level control and local-level control. The corridor level serves to act in a supervisory capacity to dynamically allocate traffic among the freeways, frontage roads, and signalized arterials. The local level of control serves to optimize flow over these individual facilities. The corridor-level process is performed periodically, typically every 10-15 minutes, or whenever measured corridor conditions appear to warrant immediate optimization. The corridor level algorithm employs first a dynamic Origin Destination algorithm that determines the entry-exit travel patterns of the freeway motorists. The estimates of O-D are based on a combination of (1) historical demand data, and (2) synthesized O-D data, generated from estimated on- and off-flows at system entrance and exit ramps. A diversion algorithm that has as its objective function any combination of the following states is then employed to produce a set of optimal diversion fractions at each control node: travel time, throughput, speed, delay, fuel consumption, and pollutant emissions. The computation of the optimal diversion fractions involves the computation of a least-cost path calculation, an alpha search process, and a traffic prediction step.

Once the optimal diversion fractions have been estimated, the next stage in the framework is to display diversion messages. A crucial element in the framework is the employment of a traffic prediction module that predicts the future state of the system for the diversion fractions selected. Traffic prediction is accomplished using the hydrodynamic theory of traffic flow for limited access routes. For signalized arterials, the extra flow due to diversion is taken into account in determining the objective function. The link costs for each link are computed using pre-simulated values of link costs for given link flow. The framework utilizes the predicted states of the system in determining arterial signalization control and ramp metering control strategies. The present state of the system as measured by traffic sensors is used as a feedback to these control strategies.

This framework recognizes and emphasizes the completion of the feedback loop between the system outputs and the control inputs. The division into control hierarchy is a very interesting and appealing feature of this framework. The provision on multiple objective functions in the diversion algorithm incorporates flexibility in determining the appropriate diversion policy. Berger et al. [8] also presents a diversion model. The paper presents the development and application of diversion control policies for the routing of freeway traffic at a single point. The diversion control algorithm uses a switching plane to compare the relative performance of each roadway defined by each of the criterion stated below:

(1) Delay difference: Control based on difference in travel delay experienced by the inter-city motorist on each roadway,
(2) Delay rate difference: Control based on an equalization of the rate of increase of motorist travel delay on each roadway,
(3) Total roadway delay difference: Control based on the difference in system delay of all motorists on each roadway.

The comparison is made by defining the appropriate switching boundaries for taking control actions. Therefore, these control algorithms are on-off types of controllers. The Sperry traffic analysis and simulation program is used for system evaluation. The simulation studies showed that the control algorithms were able to substantially reduce delays and improve traffic flows.

Zoe Ketselidou [9] has developed an expert system model for diverting traffic flow for post-incident traffic control. The model developed is based on pre-determined weights assigned to links based on the time of day and the historical traffic volumes. Points at which diversion can be initiated and the potential destinations for particular links are also determined beforehand. The explicit consideration of factors other than traffic volumes in the determination of diversion routes is the salient feature of this model.

The research at the Center for Transportation Research at Virginia Tech is an effort to analyze incident conditions so as to develop accurate diversion and re-routing strategies. At the present time, an elaborate framework has been defined and several components have been developed [10, 11]. The DTA/DTR controllers discussed in this book were designed as part of the overall framework designed to alleviate congestion caused by incidents.

# 4. Feedback Control

To understand feedback control, let us discuss the signal timing control problem. A pretimed signal timing plan has no feedback since there is no use of the real-time traffic variables. Time of day signal timing plans also do not fall into the category of real-time feedback control even if the decisions for which traffic plans to use might be dependent upon average traffic conditions. Although in this case there is a feedback loop, the loop is not closed at each sample interval when data is available. A real-time signal timing control would be the case when traffic data from sensors such as cameras or loop detectors at every sample time (or some time interval comparable to that) is input to a processor which then makes immediate decisions about affecting the traffic intersection control signals. This is very similar to the case when a police officer directs traffic at an intersection with his hands after looking at the queues at each time and deciding which stream of traffic to block and which stream to let go. The police officer's eyes are do the work of traffic sensors, his or her brain does the job of the processor, and his or her hands do the job of signal lights.

The topic of control systems deals with dynamic systems that can be controlled by some variables to produce a desired system behavior. There are two types of control systems: open loop and closed loop. Open loop systems are generally used for planning kinds of problems, where one needs to determine the control values for some time interval. On the other hand, closed loop systems (also called feedback control systems) are control systems where the control variable is a function of the output of the system. For instance, we might be driving a car with the aim of maintaining a constant speed and staying in the middle of a lane. These are the two objectives of the controller. The control actuators are the steering angle and the throttle pedal. The driver (controller) of the car reads the speed of the car and if it is different than the desired speed, the driver presses or releases the pedal. Similarly, his/her steering angle also changes depending on how close he is to the center of the lane. Hence, this is a closed loop system where at each time, the control variables (pedal angle and steering angle) are decided based on the output of the system (speed and lateral deviation). This chapter is concerned with the design of feedback control systems.

Figure 1.7 shows a generic control block diagram. The model of the system to be controlled (the traffic system in this case) is represented by the "Plant" block. The plant can be modeled as linear time invariant (LTI) differential equations, linear time varying (LTV) differential equations, or nonlinear differential equations. More generally, the plant could also be represented by partial differential equations. If we remove the "feedback" connection shown, then the system will become open loop control. The controller acts on the error signal, which is generally the difference between a desired state and what is obtained from the sensor data. There are also some controllers that are adaptive. The adaptation mechanism allows for real-time tuning of some controller parameters based on the input and the output of the system. In essence, the controller calculates how the input is affecting the output of the system in the feedback control loop, and then changes the control parameters in real-time to further improve the performance of the system.

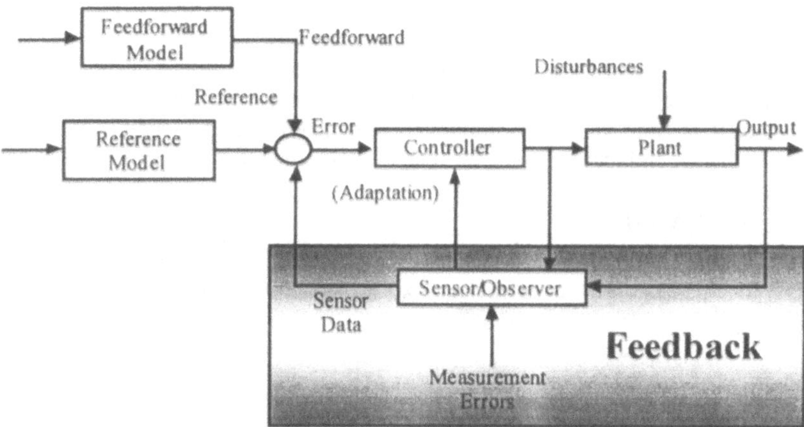


**Figure 1.7:** *Control Block Diagram*

**Control Design Steps**

The first step in control design is to come up with a mathematical model of the system to be controlled. These models are usually represented in terms of differential equations where the state variables represent various physical entities of the system. For instance, in vehicle control, the speed of the car is a state variable. The model thus obtained can be analyzed and a control law designed. The model of a system is either obtained from basic principles or by data fitting. The parameters of the model are obtained by conducting experiments on the system, collecting the data, and then performing some data fitting. Detailed knowledge of the system to be controlled is the most important aspect to a successful control design and should be given the most importance. Once the control law is designed, it can be tested in simulations with the mathematical models and then on the actual system. With the advent of microprocessors and sensor technology, most of the control implementation is based on their usage. Microprocessors are used to read the sensor values and control the actuators. The control algorithms are written in assembly language or a higher level language like C.

Control systems are generally designed using either ordinary differential equations or difference equations. The following section gives a brief introduction to both.

*Ordinary Differential Equations*

The relation between a variable and its derivatives with respect to an independent variable is called an Ordinary Differential Equation (ODE). In control theory problems, the independent variable is time. An example of an ODE is

$$\dddot{x} + 2t\dot{x} - 3\ddot{x}^2 = 0 \tag{1}$$

The highest degree of the dependent variable (x) defines the degree of the system. Equation (1) has degree 3. It is a nonlinear ODE because of the presence of the term $\ddot{x}^2$, and it is time varying (inhomogeneous) due to the presence of the time variable t in the second term. This ODE can be represented in a vector differential equation form

$$\dot{\mathbf{x}}(t) = \mathbf{f}(t, \mathbf{x}(t)) \tag{2}$$

where x is the state variable vector, $\mathbf{x} = [x^1, x^2, \ldots, x_n]^T$, and n is the system order, which is 3 for this case. Let us define $x_1 = x$, $x_2 = \dot{x}$, and $x_3 = \ddot{x}$, so that we get

$$\begin{aligned} \dot{x}_1 &= x_2 \\ \dot{x}_2 &= x_3 \\ \dot{x}_3 &= -2tx_2 + 3x_3^2 \end{aligned} \tag{3}$$

which is of the form (2) with $\mathbf{f}(t, \mathbf{x}(t)) = [x_2, x_3, -2tx_2 + 3x_3^2]^T$. In general, if f(t,x(t)) is linear in x, it is called a linear system; otherwise, it is nonlinear. Moreover, if it is independent of t, it is called time invariant; otherwise, it is called time varying. The initial value problem for an ODE is the problem of finding the value of x(t) for all future time, when system (2) is given with the value of x at initial time.

*Difference Equations*

The general form of a vector difference equation is similar to (2) and is given by

$$x(k+1) = f(k, x(k)) \tag{4}$$

where k is sample time instant. Instead of derivatives in continuous times, we have time incremented terms such as x(k+1), x(k+2), etc. Difference equations are classified the same way as are ODEs. For difference equations, we are also interested in initial value problems.

**Feedback Control Example**

Mathematically, feedback control laws are designed so that the control variable is a function of the sensed variables, and the error variable goes to zero in time. Let us consider a simple model for the longitudinal cruise control for a car as shown in Figure 1.8.

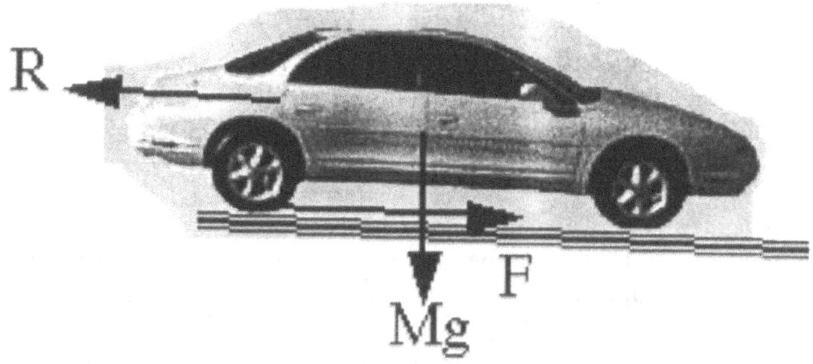


**Figure 1.8:** *Cruise Control Example*

The tractive force driving the car is given by F, which is the control variable (and can be controlled by the throttle angle in the engine intake manifold). R represents all the resistance to the car longitudinal motion and includes the air drag and other resistances like rolling and grade resistances. R in general is dependent on the car speed. M is the mass of the vehicle and g is the acceleration due to gravity. If we denote x as the velocity of the car (state variable for the system) and $x_d$ the desired speeds, then we can write the differential equation for the system as

$$\dot{x}(t) = -\frac{R(x,t)}{M} + \frac{F(t)}{M} \tag{5}$$

Our aim is to design a control law that will take the error variable defined as $e(t) = x(t) - x_d(t)$ to go to zero. Since the desired cruise speed is constant, we have

$$\dot{x}_d(t) = 0 \tag{6}$$

Subtracting (6) from yields

$$\dot{e}(t) = -\frac{R(x,t)}{M} + \frac{F(t)}{M} \tag{7}$$

Now, if we take the control law as

$$F(t) = M[\frac{R(x,t)}{M} - ke(t)] \tag{8}$$

then by substituting this equation in (7) we obtain the closed loop dynamics of the system as

$$\dot{e}(t) = -ke(t) \tag{9}$$

The solution of this differential equation is easily obtained by integration and is given by

$$e(t) = e(o)e^{-kt} \tag{10}$$

This shows that e(t) will go exponentially to zero if we choose k to be a positive constant in the control law (8).

This example was easy, but in general, there are many complications in designing control systems. For example, usually the problem will have a higher order than this system. The number of control variables can also be more than one. There can be uncertainties in the system, implying that the controller does not have exact knowledge of the parameters such as M, R, etc.

**Other Issues**

Feedback control for DTR can be an effective solution for alleviating traffic congestion during major incidents. However, the success of such a system depends on the effective modeling of the system as well as the design of the appropriate control law. The designer of the controller needs to address issues such as controllability and observability of the traffic system, actuation and sensing, robustness, and stability of the closed loop system.

*Actuation and Sensing*

The actuation of this system can be achieved in many ways, such as Variable Message Sign (VMS), in-vehicle guidance, and Highway Advisory Radio (HAR). State variables such as the traffic density and average traffic speed. can be sensed using various types of traffic sensors such as inductive loops, traffic cameras, and transponders.

*Controllability and Observability*

The designer should analyze the system before designing the controller to determine if the system is controllable and observable. Controllability implies that a suitable control law can be devised in order to obtain a desired response from the system. Control action in traffic control is advisory, so in the system description, compliance of the drivers should also be modeled to evaluate the controllability of the overall system. Compliance in general also varies with time and is a function of many variables. No analytic models of driver compliance have been developed yet, so researchers usually test the controllers using fixed partial compliance values. Observability implies that the system state variables can be observed from the sensed output. For instance, if the system is not controllable, then we might decide to add more actuation infrastructure such as VMS, HAR, etc., and if the system is not observable, we might add more sensors to the system.

*Robustness and Stability*

The effectiveness of the control design can be measured in terms of its robustness, stability, and transient characteristics. A robust controller will perform well even in the presence of uncertainties in the nominal model of the system. Models representing traffic systems cannot represent the system fully, and therefore there are uncertainties in the system which have to be addressed. A control law should provide stability to the system and desirable transient response. For instance, a good DTR control law would minimize time for the system to change from a congested state to a normal flow state. Stability in the traffic control sense would imply that starting from a given difference in travel times on the alternate routes, the controller is able to keep all future travel time differences within some bound.

## 5. Summary

Dynamic Traffic Routing is important for Traffic Congestion Management
It can be used for distributing traffic at traffic nodes so that either the travel time on alternate routes is equated or the total travel time on all routes is minimized.
The overall system consists of traffic sensors, a computer for processing the sensed traffic variables, and traffic actuators such as VMS, lane control signs, HAR, signal timing control, and ramp control.
Feedback control is very promising since it performs measurements in real-time to calculate the control measures utilizing the sensed data in a closed-loop setting which is traffic responsive, and can be designed to be robust against disturbances.

## 6. Exercises

### Questions

Question 1: What is dynamic traffic routing and why is it important?
Question 2: What are the two objectives that a dynamic traffic assignment algorithm can satisfy?
Question 3: What is the difference between an open-loop and closed-loop traffic control?
Question 4: What is feedback control, and how can it be applied to dynamic traffic assignment?
Question 5: What kinds of sensors are used for traffic surveillance, and what traffic variables do they sense?
Question 6: What are some mechanisms to control traffic behavior?

### Problems

*Problem 1*

Consider the following two-node network (Figure 1.9). Assume that the traffic density on route one is $\rho_1 = 3$ vehicles, and on the second route is $\rho^2 = 6$ vehicles. We have u = 5 vehicles which need to be distributed within the two routes in the next sample time, and we assume that no vehicles left the two routes during that time. Assume that travel time on each route is given by

$$T_i = \frac{d_i}{v_{fi}(1 - \rho_i / \rho_{mi})}$$

where:

| $T_i$ | Travel time on route i |
|---|---|
| $d_i$ | Length of route i |
| $v_{fi}$ | Freeflow speed on route i |

| | |
|---|---|
| $\rho_{mi}$ | Jam density on route i |
| $\rho_i$ | Traffic density on route i |

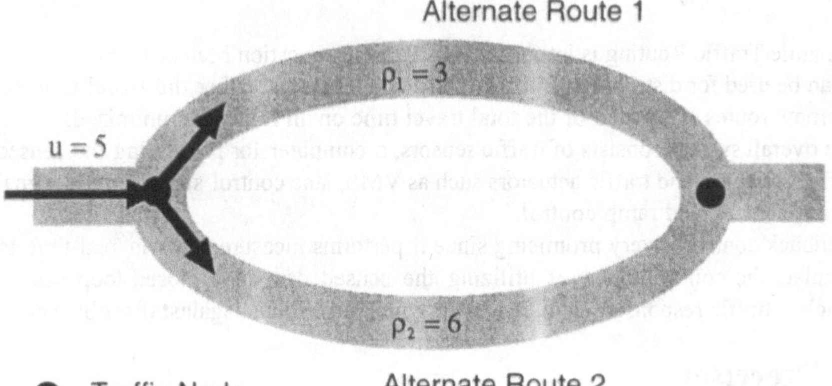


**Figure 1.9:** *Problem Network for Point Diversion*

Assume for this problem that both routes have the same length, jam densities, and free flow speeds. Let $u_1$ be the number of vehicles out of 5 that you will assign to route 1, and let $u_2$ be the number of vehicles assigned to route 2. Notice that we should have $u_1 + u_2 = 5$. Find $u_1$ and $u_2$ so that:

a) The difference between the travel times on the two routes is minimized (user-equilibrium)
b) Total travel time on the two routes is minimized (system optimal)

*Problem 2*

Now consider Figure 1.10. Here we have traffic coming at node 1 destined towards node 2 and node 3. There are two alternate routes to go from node 1 and node 2 (one route on link 1, and the other route which consists of links 2 and 3), and two alternate routes from node 1 to node 3 (one route on link 2, and the other route which consists of links 1 and 3).

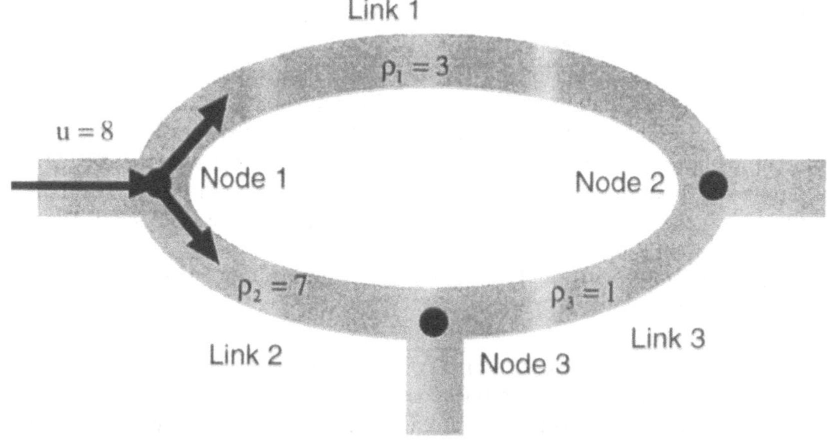


**Figure 1.10:** *Problem Network for Traffic Assignment*

Now assuming the same assumptions of problem 1, calculate $u_1$ and $u_2$ so that:

(4) The difference between the travel times on the two sets of alternate routes is minimized (user-equilibrium)

(5) Total travel time on all the links is minimized (system optimal)

*Problem 3*

In the cruise control example, if the value of the control gain k is 15, how much time will it take for the initial speed error to become half its value?

*Problem 4*

Write a computer program to simulate the cruise control example. Use M=1000 Kg, g=9.8 m/$S^2$, and R= 0.5 $x^2$+0.04xM/1000. Take the desired speed of the car to be 75 Km/h. Plot speed versus time when the initial speed of the car is 65 Km/h. Note that you will need a discretized version of the differential equation (7). You can either take first order discretization (so that $\dot{x} = [x(k+1) - x(k)]/h$, where k is the discrete time step and h is chosen to be appropriately small), or use a higher order method such as the Runga-Kutta algorithm (see any book on numerical algorithms).

## 7. References

(1) Lighthill M J, Whitham G B 1955 On kinematic waves II. A theory of traffic flow on long crowded roads. Proc of the Royal Society of London, Series A 229: 317-345

(2) Papageorgiou M, Messmer A 1991 Dynamic network traffic assignment and route guidance via feedback regulation. Transportation Research Board, Washington, D.C.

(3) Papageorgiou M 1990 Dynamic modeling, assignment and route guidance traffic networks. Transportation Research 24B: 471-495
(4) Messmer A, Papageorgiou M 1995 Route diversion control in motorway networks via nonlinear optimization. IEEE Trans on Control Systems Tech 3:
(5) Messmer A, Papageorgiou M 1994 Optimal freeway network control via route recommendation. Vehicle Navigation & Information Systems Conference Proc
(6) Gupta A, Maslanka V, Spring G 1992 Development of a prototype KBES in the management of non-recurrent congestion on the Massachusetts Turnpike. 71st Annual TRB
(7) Gartner N H, Reiss R A 1987 Congestion control in freeway corridors: the IMIS system. NATO ASI Series F38, Flow Control of Congested Networks, Springer-Verlag
(8) Berger C R, Gordon R L, Young, P E 1976 Single point diversion of freeway traffic. Transportation Research Record 601: 10-17
(9) Ketselidou Z 1993 Potential use of knowledge-based expert system for freeway incident management. ITS, University of California, Irvine
(10) Özbay K, Hobeika A G, Subramaniam S, Krishnaswamy V 1994 A heuristic network generator for traffic diversion during non-recurrent congestion. Transportation Research Board 73rd Annual Meeting, Washington, D.C.
(11) Hobeika A G, Sivanandan R, Subramaniam S, Özbay K, 1993 Real-time traffic diversion model: conceptual approach. ASCE J of Transp 119: 515-534

# CHAPTER 2
# TRAFFIC FLOW THEORY

## Objectives

To present the basic traffic flow theory which will be used in the next chapter for control problem formulations.

## 1. Introduction

To design traffic controllers, it is very important to understand the basics of traffic flow theory. This chapter presents the relevant fundamentals of traffic flow theory. Traffic flow theory is mainly presented from the macroscopic perspective where aggregate traffic variables, such as traffic flow, traffic density, and average traffic speed, are considered. The macroscopic characteristics are also related to the microscopic behavior of traffic.

## 2. Conservation Law

The conservation law in the case of traffic problem can be stated as "vehicles can neither be created nor destroyed on a highway section". That implies that number of vehicles coming in and going out of a highway section account for the change of traffic density on that section. Let us consider a section of the highway at distance x meters from some reference point and consider the section of length $\Delta x$ as shown in Figure 2-1.

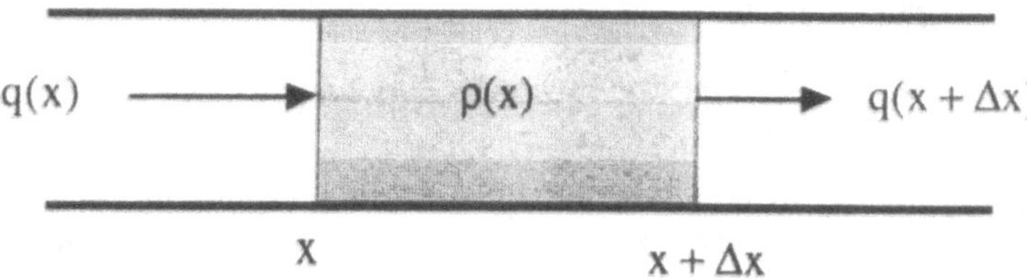


**Figure 2.1:** *Conservation Equation for Traffic*

Now, the change in number of vehicles in the section of length $\Delta x$ is given by

$$\frac{d}{dt}\int_{x}^{x+\Delta x}\rho(\ell,t)d\ell \tag{1}$$

The change in the number of cars in this section is also equal to the number of cars going out of this section subtracted from the number of vehicles coming in, as shown below.

$$\frac{d}{dt}\int_{x}^{x+\Delta x} \rho(\ell,t)d\ell = q(x,t) - q(x+\Delta x) \tag{2}$$

Now, we can use the fundamental theorem of calculus to rewrite equation (2) as

$$\int_{x}^{x+\Delta x} \frac{\partial \rho(\ell,t)}{\partial t} d\ell = -\int_{x}^{x+\Delta x} \frac{\partial q(\ell,t)}{\partial \ell} d\ell \tag{3}$$

Note that instead of the interval x and $x+\Delta x$, we could have used any interval from a to b, and the same analysis would work. Now, we can take $\Delta x \to 0$ in the next step if we are using $\Delta x$ or if we use any arbitrary interval, then we can observe that since the result is true for any arbitrary a and b, we obtain the mass conservation model of a highway, characterized by x $\in$ [0, L], which is the position on the highway, as

$$\frac{\partial}{\partial t}\rho(x,t) = -\frac{\partial}{\partial x} q(x,t) \tag{3}$$

where $\rho(x,t)$ is the density of the traffic as a function of x, and time t, and $q(x,t)$ is the flow at given x, and t. The flow $q(x,t)$ is a function of $\rho(x,t)$, and the speed v(x,t), as shown below:

$$q(x,t) = \rho(x,t)\,v(x,t) \tag{4}$$

This model of a highway section is shown in Figure 2.2.

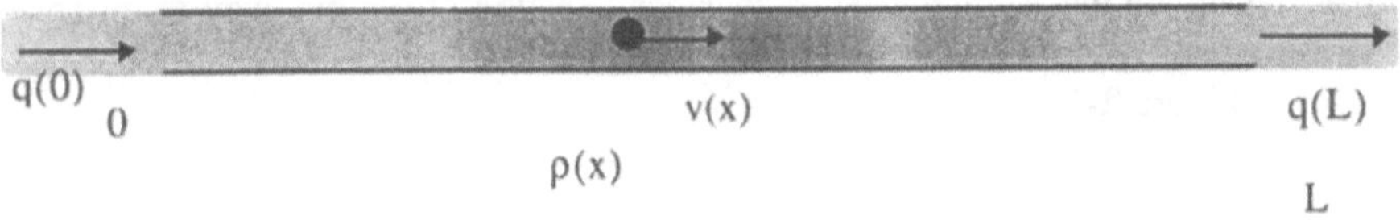


**Figure 2.2:** ***Segment of Highway Model***

# 3. Traffic Density-Flow Relationships

Many models have been proposed to represent the relationship between traffic density and traffic flow. The following is a brief description of some of these models

**Greenshield's Model [1]**

In Greenshield's model, the speed-density relationship is a linear relationship given by

$$v = v_f\left(1 - \frac{\rho}{\rho_{max}}\right) \tag{5}$$

Traffic flow, traffic speed, and traffic density have a fundamental relationship which is true in any model. These variables are related as:

$$q = v\rho \tag{6}$$

Therefore, for Greenshield's model

$$q = \rho v_f (1 - \frac{\rho}{\rho_{max}}) \tag{7}$$

The relationships between the three variables are shown in Figure 2.3. In the first plot in the figure, we see that the slope of the flow-density relationship is equal to the free flow speed. This can be shown by differentiating (5) to obtain

$$\frac{dq}{d\rho} = v_f - 2v_f \frac{\rho}{\rho_{max}} \tag{8}$$

The value of this slope at $\rho = 0$ is $v_f$. We can obtain the maximum flow $q_{max}$ by equating (5) to 0. That gives us the value of the density $\rho_c$ at maximum flow as

$$\rho_c = \frac{\rho_{max}}{2} \tag{9}$$

Plugging this value of density in A.3 yields

$$q_{max} = \frac{v_f \rho_{max}}{4} \tag{10}$$

Figure 2.3 also shows the congested and uncongested region of traffic flow.

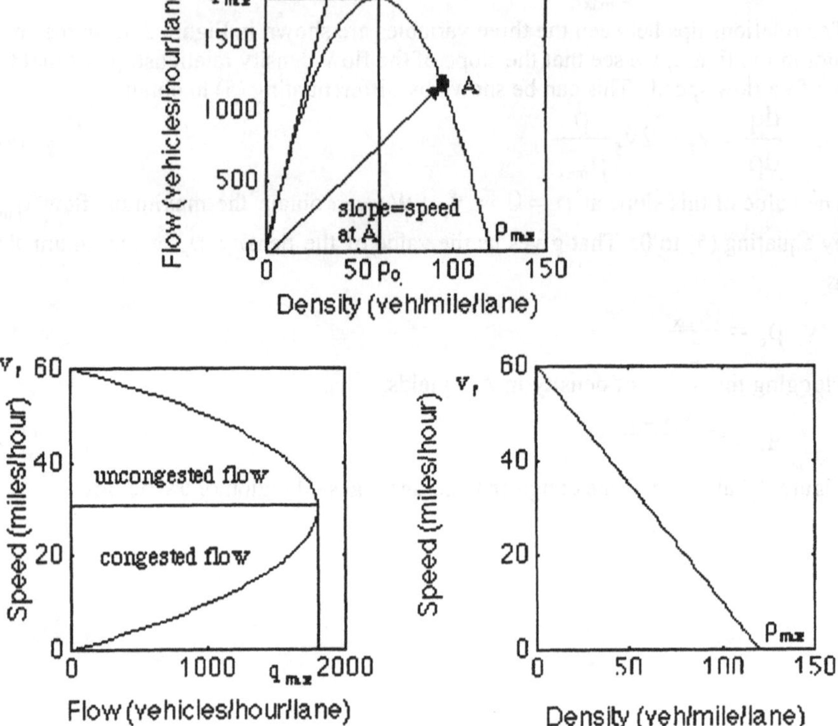


**Figure 2.3:** *Traffic Flow Diagrams*

**Greenberg Model [2]:**

In the Greenberg model, the speed-density relationship is given by

$$v = v_0 \ln\left(\frac{\rho}{\rho_{max}}\right) \tag{11}$$

**Underwood Model [3]:**

In the Underwood model, the speed-density relationship is given by

$$v = v_f e^{-\rho/\rho_0} \tag{12}$$

**Northwestern University Model [4]:**

In the Northwestern University model, the speed-density relationship is given by

$$v = v_f e^{-1/2(\rho/\rho_0)^2} \tag{13}$$

**Drew Model [5]:**

In the Drew model, the speed-density relationship is given by

$$v = v_f \left[ 1 - \left( \frac{\rho}{\rho_{max}} \right)^{n+1/2} \right] \tag{14}$$

Drew's model is a generalization of earliar models, where taking the value of n=1 gives a the linear model, n=0 gives a parabolic model, and n=-1 gives an exponential model.

**Pipes-Munjal Model [6]:**

In the Pipes-Munjal model, the speed-density relationship is given by

$$v = v_f \left[ 1 - \left( \frac{\rho}{\rho_{max}} \right)^{n} \right] \tag{15}$$

The Pipes-Munjal model, like Drew's model, also is a generalization where by taking different values of n, we obtain different models.

**Multiregime Models [7]:**

In multiregime models, different regions of traffic are modeled by different models. For example, congested regions and uncongested regions might use different models.

**Diffusion Models:**

In order to account for the fact that drivers look ahead and modify their speeds accordingly, (5) can be replaced by

$$v_e = v_f (1 - \frac{\rho}{\rho_{max}}) - D(\frac{\partial \rho}{\partial x})/\rho \tag{16}$$

Using (16) and the fact $q = \rho\, v_e$, relationship (4) now can be replaced by

$$q(x,t) = \rho(x,t)\, v(x,t) - D \frac{\partial}{\partial x} \rho(x,t) \tag{17}$$

where D is a diffusion coefficient given by

$$D = \tau v_r^2 \tag{18}$$

where $v_r$ is a random velocity, and $\tau$ is a parameter. Diffusion is a useful concept mentioned by many researchers as an extension to the existing traffic flow models to improve their realism [8,9,10]. Diffusion term represents "the diffusion effect" due to the fact that each driver's gaze is concentrated on the road in front of him/her, so that he/she adjusts his/her speed according to the concentration ahead. This adjustment creates a dependence of flow on concentration gradient which leads to an effective diffusion. This models the gradual rather than instantaneous reduction of speed by the drivers in response to the shock waves. Combining equations (3) and (17) gives

$$[\frac{\partial}{\partial t}\rho(x,t) + v_f \frac{\partial}{\partial x}\rho(x,t)] - 2\frac{\rho}{\rho_{max}} v_f \frac{\partial}{\partial x}\rho(x,t) - D\frac{\partial^2}{\partial x^2}\rho(x,t) = 0 \quad (19)$$

If we introduce a moving reference frame

$$\xi(x,t) = -x + v_f t \quad (20)$$

and non-dimensionalize $\rho(x,t)$ by $\rho_{max}/2$, and t by $t_0$, equation (7) gets transformed to

$$\frac{\partial}{\partial t}\rho(\xi,t) + \rho\frac{\partial}{\partial \xi}\rho(\xi,t) - \frac{1}{R_e}\frac{\partial^2}{\partial \xi^2}\rho(\xi,t) = 0 \quad (21)$$

Here, $R_e$ is a dimensionless constant, and is analogous to the Reynolds number in fluid dynamics. $R_e$ is given by

$$R_e = (\frac{v_f}{v_r})^2 \frac{t_0}{\tau} \quad (22)$$

Equation (21) shows the Burgers' equation formulation of the traffic flow problem. Some researchers have also worked on the conservation law

$$\frac{\partial}{\partial t}\rho(x,t) + \rho(x,t)\frac{\partial}{\partial x}\rho(x,t) = \varepsilon\frac{\partial^2}{\partial x^2}\rho(x,t) \quad (23)$$

with a solution obtained by taking the following limit.

$$\rho(x,t) = \lim_{\varepsilon \to 0} \rho^{\varepsilon}(x,t) \quad (24)$$

where $\rho^{\varepsilon}(x,t)$ satisfies (21) [11-16]. Using this form reduces the Burgers' equation formulation into the classical traffic model with no diffusion.

Note that instead of adding a diffusion term to (5) we could have started from a different static model and introduced a diffusion term to that.

# 4 Microscopic Traffic Characteristics [17,18,19]

Macroscopic traffic dynamics represent traffic in terms of traffic density, flow and spped. We can also view traffic in terms of its microscopic characteristics, i.e. studying individual vehicle behavior, Fortunately there is a link between the two as presented below.

Microscopic traffic characteristics are described in terms of the car-following models [7,18]. The car-following model is developed based on Figure 2.4. The vertical double line is a reference from where the x-axis distance for cars is calculated. We show the situation where car n+1 is following car n. L is the distance between the two cars at rest and is a constant. The variable h is the headway distance between the cars. The car-following model is based on assumptions of how human drivers vary h as a function of other variables.

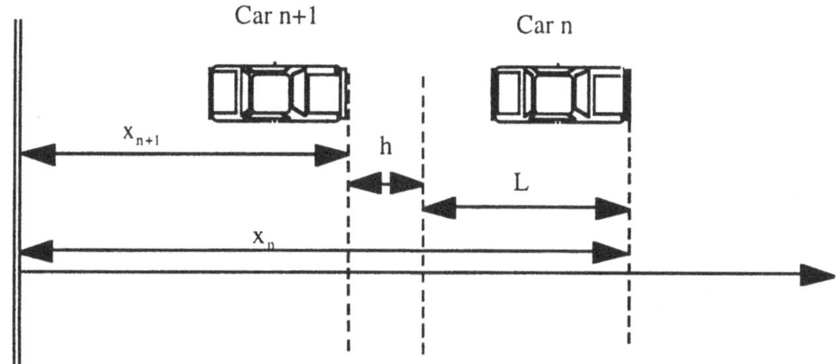


**Figure 2.4:** *Car-Following Model*

For example, if we make h a linear function of the speed of the follower, we obtain

$$x_n - x_{n+1} = k\dot{x}_{n+1} + L \tag{25}$$

Here k and L also are constants. By differentiating (25) we get

$$\ddot{x}_{n+1} = \frac{1}{k}[\dot{x}_n - \dot{x}_{n+1}] \tag{26}$$

This model is enhanced by introducing a driver delay $\tau$ to the stimulus provided by the leader car. We also replace (1/k) by another variable $\lambda$ called the sensitivity. Using this, (26) takes the form

$$\ddot{x}_{n+1}(t+\tau) = \lambda[\dot{x}_n(t) - \dot{x}_{n+1}(t)] \tag{27}$$

where

$$\lambda = \lambda_0 \frac{[\dot{x}_{n+1}(t+\tau)]^m}{[x_n(t) - x_{n+1}(t)]^L} \tag{28}$$

Here, m, and L are integer values and $\lambda_0$ is a constant.

It is very remarkable that we can obtain the macroscopic models by choosing different values for m and L and then integrate Equation (27). For Greenshield's model m=0, and L=2 and for Greenberg model, m=0 and L=1. For example, let us take m=0 and L=2 in (26). We get

$$\ddot{x}_{n+1}(t+\tau) = \lambda_0 \frac{[\dot{x}_n(t) - \dot{x}_{n+1}(t)]}{[x_n(t) - x_{n+1}(t)]^2} \tag{29}$$

Integrating the above, we obtain

$$\dot{x}_{n+1}(t+\tau) = -\frac{\lambda_0}{x_n(t) - x_{n+1}(t)} + C \tag{30}$$

where C is a constant of integration. We use the relationship between average space headway and traffic density as

$$x_n - x_{n+1} = \frac{1}{\rho} \tag{31}$$

and also consider steady state conditions so that

$$\dot{x}_{n+1}(t+\tau) = \dot{x}_n(t) = v \tag{32}$$

Using (31) and (32) in (30), we obtain

$$v = -\lambda_0 \rho + C \tag{33}$$

By using the boundary conditions such that v= $v_f$ at $\rho$=0 and v=0 at $\rho$= $\rho_{max}$, we obtain

$$v = v_f\left(1 - \frac{\rho}{\rho_{max}}\right) \tag{34}$$

# 5 Traffic Model

We will use the following notation from [20].

$$u_t = \frac{\partial u}{\partial t} \quad u_x = \frac{\partial u}{\partial x} \quad u_{xx} = \frac{\partial^2 u}{\partial x^2} \ldots$$

The traffic problem in the PDE (Partial Differential Equations) setting is given by

$$\frac{\partial}{\partial t}\rho(x,t) = -\frac{\partial}{\partial x}q(x,t) \quad -\infty < x < \infty \quad 0 < t < \infty \tag{35}$$

with the initial condition

$$\rho(x,0) = \phi(x) \quad -\infty < x < \infty \tag{36}$$

We can write

$$\frac{\partial q}{\partial x} = \frac{\partial q}{\partial \rho}\frac{\partial \rho}{\partial x} \tag{37}$$

or

$$\frac{\partial q}{\partial x} = c(\rho)\frac{\partial \rho}{\partial x} \tag{38}$$

where $c(\rho) = \dfrac{\partial q}{\partial \rho}$ (39)

Substituting this in (35) and using the notation (34) we get the representation of the traffic problem as

PDE $\quad \rho_t + c(\rho)\rho_x = 0 \quad -\infty < x < \infty \quad 0 < t < \infty$ (40)

IC $\quad \rho(x,0) = \phi(x) \quad -\infty < x < \infty$ (41)

Here, IC stands for initial condition.

In the model (40-41) we are using an infinite length highway. If we want to model a finite length model with length L, we would use the following model

PDE $\quad \rho_t + c(\rho)\rho_x = 0 \quad 0 \le x \le L \quad 0 < t < \infty$ (42)

IC $\quad \rho(x,0) = \phi(x) \quad 0 \le x \le L$ (43)

BC $\quad \rho(0,t) = \psi(t) \quad 0 < t < \infty$ (44)

Here, BC stands for boundary condition, which dictates how the traffic is entering the highway section.

# 6 Classification of PDEs [20,21,22]

In this section we will classify the traffic problem. In order to do that, we first present the basic information on PDEs.

### Variables

In (40-41), $\rho$ is the *dependent variable,* which depends on the two *independent variables* x and t. We can have *systems of PDEs* which contain more than one dependent variables. We will encounter problems like this in traffic routing.

### Order of the PDE

Order of a PDE is the order of the highest partial derivative. In (40), the order of the PDE is one.

### Linearity

PDEs are linear if the dependent variable with all its derivatives appears linearly in the equation. For example, a second order linear PDE in two independent variables x and y is

$$Au_{xx} + Bu_{xy} + Cu_{yy} + Du_x + Eu_y + Fu = G \tag{45}$$

where A,B,C,D,E,F, and G are either *constants* or functions of the independent variables x and y (but not u or any of its derivatives). If G(x,y) is identically zero, then (45) is called *homogeneous*, otherwise it is *nonhomogeneous*. If a PDE is not linear then it is *nonlinear*. If any of A,B,C,D,E,F, or G is a function of u or any of its derivatives, then this equation is called *quasilinear*. If on;ly G is a function of u , and if A,B,C,D,E,F are independent of u, then this will be called *almost linear* PDE. If only Linear second order PDEs like (45) are of the following three kinds.

Parabolic If a PDE like (45) satisfies $B^2 - 4AC = 0$, then it is parabolic. Heat flow and diffusion equations are of this type, e.g. $u_t = u_{xx}$.

Hyperbolic: If a PDE like (45) satisfies $B^2 - 4AC > 0$, then it is hyperbolic. Vibrating systems and wave motion equations are of this type, e.g. $u_{tt} = u_{xx}$.

Elliptic: If a PDE like (45) satisfies $B^2 - 4AC < 0$, then it is elliptic. Laplace's equation is an example of this kind, e.g. $u_{xx} + u_{yy} = 0$.

The traffic problem (40) is a nonlinear first order PDE with two independent variables and one dependent variable. It is also classified as a quasi-linear PDE since the nonlinear element c( $\rho$ ) enters linearly in the equation.

**Boundary Conditions**

There are three types of boundary conditions. In one condition (*Dirichlet boundary condition* or the boundary condition of the first kind), the value of the dependent variable is specified, in the second kind *(Neumann boundary condition* or the boundary condition of the second kind), the gradient of the dependent variable is specified, and in the third kind *Robin boundary condition* or the boundary condition of the third kind), a sum of first two is specified. For example

$$u(0,t) = g_1(t) \tag{46}$$

$$u(L,t) = g_2(t)$$

is a Dirichlet boundary condition.

$$u_x(0,t) = k_1(t) \tag{47}$$

$$u_x(L,t) = k_2(t)$$

is a Neumann boundary condition.

$$u(0,t) + c_1 u_x(0,t) = h_1(t) \tag{48}$$

$$u(L,t) + c_2 u_x(L,t) = h_2(t)$$

is a Robin boundary condition.

# 7 Existence of Solution [21]

After the physical system like the traffic problem is modeled, the well-posedness of the model is to be addressed. A problem is well posed if :

1. A solution of the problem exists
2. The solution is unique
3. The solution depends continuously on the data of the problem

The existence of a solution for the traffic problem can be studied by analyzing the following PDE.

$$P(t,x,\rho)\rho_t + Q(t,x,\rho)\rho_x = R(t,x,\rho) \tag{49}$$

where P, Q and R are functions defined and $C^1$ in some domain belonging to $R^3$. $C^1$ functions are those functions which are continious with continuous partial derivatives upto order 1. The solution of this problem is a function if it exists $\rho(t,x)$ defined and $C^1$ in some domain belonging to $R^2$, so that when it is substituted in (49), it gives identity for all (t,x) in the domain.

Let us consider *characteristic equation* for

$$u(t,x,\rho) = 0 \tag{50}$$

Let us parameterize the curve generated by the solutions as:

$$t = t(\tau) \quad x = x(\tau) \quad \rho = \rho(\tau) \tag{51}$$

Since u is a constant, we obtain the following

$$\frac{d}{dt}u(t(\tau),x(\tau),\rho(\tau)) = \frac{\partial u}{\partial t}\frac{\partial t}{\partial \tau} + \frac{\partial u}{\partial x}\frac{\partial x}{\partial \tau} + \frac{\partial u}{\partial \rho}\frac{\partial \rho}{\partial \tau} \tag{52}$$

$$= \frac{\partial u}{\partial t}P + \frac{\partial u}{\partial x}Q + \frac{\partial u}{\partial \rho}R = 0$$

where

$$P = \frac{\partial t}{\partial \tau} \quad Q = \frac{\partial x}{\partial \tau} \quad R = \frac{\partial \rho}{\partial \tau} \tag{53}$$

We can solve the problem by using (53) as

$$\frac{dt}{P} = \frac{dx}{Q} = \frac{d\rho}{R} \tag{54}$$

Applying implicit function theorem[1] on (52) gives

$$\rho_t = -\frac{u_t}{u_\rho}, \quad \rho_x = -\frac{u_x}{u_\rho} \tag{53}$$

Therefore,

$$P\rho_t + Q\rho_x = -\frac{Pu_t + Qu_x}{u_\rho} = \frac{Ru_z}{u_z} = R \tag{54}$$

which shows that the characteristic equation method does solve (49).

**Traffic Problem**

Let us analyze the following traffic problem for existence of solution from equations (42-43).

$$\rho_t + c(\rho)\rho_x = 0 \tag{55}$$

$$\text{IC} \quad \rho(x,0) = \phi(x) \quad 0 \le x \le L \tag{56}$$

To find the solution, we use (54) to obtain

$$\frac{dt}{1} = \frac{dx}{c(\rho)} = \frac{d\rho}{0} \tag{57}$$

Solving these ode's (ordinary differential equations), we get two independent solutions as

$$u_1 = \rho, \quad u_2 = x - c(\rho)t \tag{58}$$

---

[1] **Implicit Function Theorem**: Let F be a variable of three variables of class $C^1$ in an open set O given by

$$F(x,y,z)=C$$

Then z can be solved in terms of x and y for (x,y,z) near the point $(x_0,y_0,z_0)$ if $F(x_0,y_0,z_0)=0$. Moreover,

$$\frac{\partial z}{\partial x} = -\frac{F_x}{F_z}, \quad \frac{\partial z}{\partial y} = -\frac{F_y}{F_z}$$

If $u_1$ and $u_2$ are two independent solutions of the pde, then $F(u_1, u_2)$, which is an arbitrary $C^1$ function of $u_1$ and $u_2$ and provides the following *general integral* of the pde.

$$F(u_1(t,x,\rho),u_2(t,x,\rho))=0 \tag{59}$$

We can take a specific form of (59) as

$$u_2(t,x,\rho)=G(u_1(t,x,\rho)) \tag{60}$$

So, using (59) and (60) with (58) we get the following as the general integral of (55).

$$\rho=G(x-c(\rho)t) \tag{61}$$

For this solution to satisfy the initial condition (56), we obtain

$$\rho=\phi(x-c(\rho)t) \tag{62}$$

Equation (62) can be rewritten as

$$\rho-\phi(x-c(\rho)t)=0 \tag{63}$$

From implicit function theorem, the solution for $\rho$ in terms of t, x is implied by (63) if the following condition is satisfied.

$$1+c'(\rho)t\phi'(x-c(\rho)t)\neq 0 \tag{64}$$

Here the symbol ' implies that the differentiation is to be performed with respect to the variable in the parenthesis. Now at initial condition, t=0. Hence, at t=0, the left hand side of (64)>0. This implies that the solution will exist as long as left hand side of (64)>0, and as soon as it becomes equal to zero, the solution ceases to exist. When this condition is not satisfied the derivatives of the traffic density becomes infinite creating a discontinuity in the solution, which is called a traffic *shock*. This condition is restated as

$$1+c'(\rho)t\phi'(x-c(\rho)t)>0 \tag{65}$$

If we use Greenshield's formula (7), and using (39), we get

$$c(\rho)=\frac{\partial q}{\partial \rho}=v_f-2\frac{v_f}{\rho_{max}}\rho \tag{66}$$

Differentiating (66) yields

$$c'(\rho)=\frac{\partial c(\rho)}{\partial \rho}=-2\frac{v_f}{\rho_{max}} \tag{67}$$

Substituting (67) in (65) gives us

$$1-2\frac{v_f}{\rho_{max}}t\phi'(x-c(\rho)t)>0 \tag{68}$$

Notice that if $\phi'(x)\leq 0$ for all x, then (68) is always satisfied, and no shocks or discontinuities will be created for any t≥0. However, if $\phi'(x)>0$ for any distance, a shock will be developed after some time. The next section illustrates how to analyze the traffic pdes for solutions and shock waves.

# 8 Method of Characteristics to Solve First Order PDEs [20]

We can use the method of characteristics to solve a first order pde of the form

$$a(x,t)u_x + b(x,t)u_t + c(x,t)u = 0 \tag{69}$$

Notice that if we make u a function of a variable s, then

$$\frac{du}{ds} = u_x \frac{dx}{ds} + u_t \frac{dt}{ds} \tag{70}$$

If we compare (69) and (70) we can choose

$$\frac{dx}{ds} = a(x,t), \quad \frac{dt}{ds} = b(x,t) \tag{71}$$

So that the pde (69) is transformed into the ode (72).

$$\frac{du}{ds} + c(x,t)u = 0 \tag{72}$$

Which can be solved with the initial condition given in terms of another variable m. The variable m changes along the initial curve, such as the curve t=0, and the variable s will change along characteristics curve. We will notice below that for the traffic problem s will be same as t, and the variable m is the initial density on the highway which is propagated over the characteristic curves.

Comparing (69) to (55), we notice that u in (69) is same as $\rho$ and b(x,t)=1. Since, b(x,t)=1, we get from (71) that s=t, and therefore, we get again from (71)

$$\frac{dx}{dt} = c(\rho) \tag{73}$$

Since along the characteristic, $\rho$ remains constant and is equal to $\rho_0$. This method lets us follow a traffic density given at initial time. We follow the density in time and space. We can rewrite (73) as

$$\frac{dx}{dt} = c(\rho_0) \tag{74}$$

Integrating this equation, we get

$$x(t) = c(\rho_0)t + x(0) \tag{75}$$

which is a straight line with the slope $c(\rho_0)$. We can also derive these characteristics in the following way. We want to track a constant traffic density, which is given by

$$\rho(x(t),t) = \rho(x_0,0) = \rho_0 \tag{76}$$

Here the x(t) and t are the variables which show how the constant density is moving in the x-t plane. Therefore $\rho$ is constant along a curve in the x-t plane and this curve is called the *characteristic curve*. We can differentiate (76) to obtain

$$\frac{d\rho(x(t),t)}{dt} = \frac{\partial \rho(x(t),t)}{\partial t} + \frac{\partial \rho(x(t),t)}{\partial x}\frac{dx(t)}{dt} = 0 \tag{77}$$

Substituting (55) in (77) we again get (74).

The slope of the curve (75), given by $c(\rho) = \frac{\partial q}{\partial \rho}$ is called the *local wave velocity* for $\rho_0$. It shows how disturbances travel in the traffic. Notice that the disturbances can travel forward or backward in x. Obviously the local wave velocity is not the same as the average traffic velocity v which moves only forward.

**Example 1:** Let us consider the traffic system given by

$$\rho_t + c(\rho)\rho_x = 0 \tag{78}$$

Where, using Greenshield's model, we have

$$c(\rho) = \frac{\partial q}{\partial \rho} = v_f - 2\frac{v_f}{\rho_{max}}\rho \tag{79}$$

Let the initial condition be given by (Figure 2.5)

$$\rho(x,0) = \begin{cases} 1 & x \le 0 \\ 1-x & 0 < x < 1 \\ 0 & 1 \le x \end{cases} \quad -\infty < x < \infty \tag{80}$$

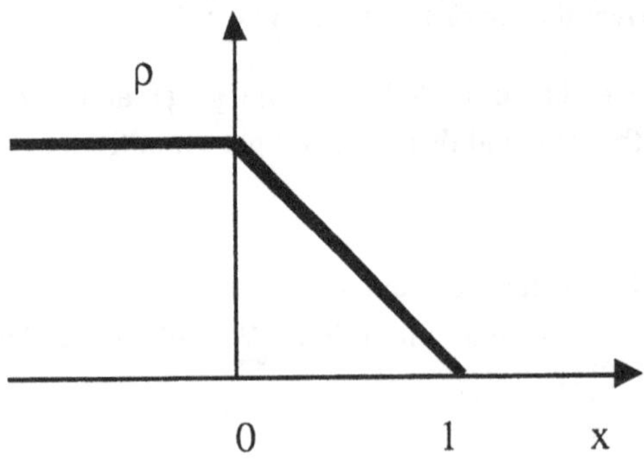


**Figure 2.5:** *Initial Traffic Density*

The slope of the characteristic curve is given by

$$c(\rho_0) = v_f - 2\frac{v_f}{\rho_m}\rho_0 \tag{81}$$

Therefore, for $\rho_0$=0, the slope is given by $v_f$. Note that for $\rho_0 < \rho_m/2$, the slope is positive and for $\rho_0 > \rho_m/2$, it is negative (showing back propagation of the wave). The propagation of the traffic densities for this problem is shown in Figure 2.6.

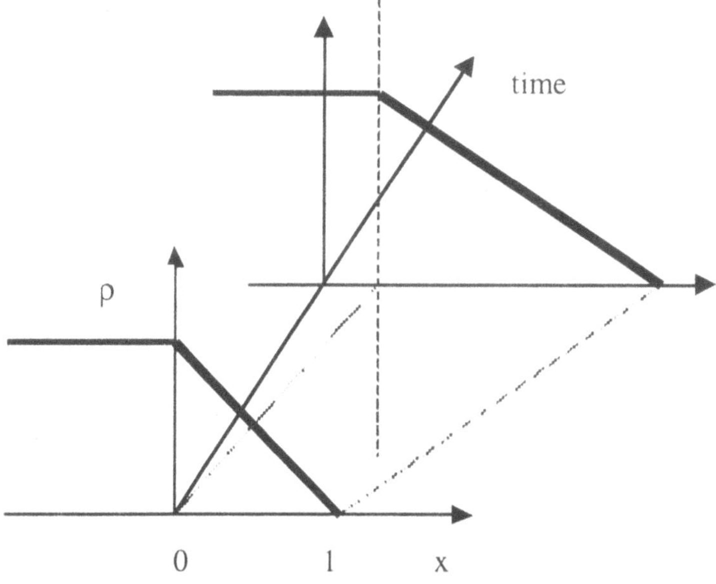


**Figure 2.6:** *Traffic Density vs. Time*

The characteristics of the same problem in the t,x plane is shown in Figure 2.7.

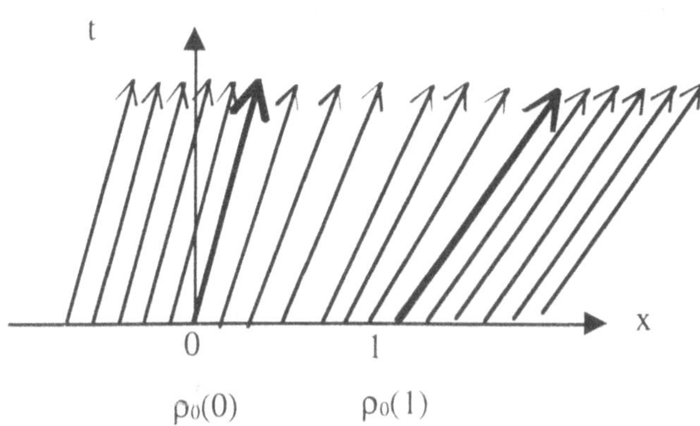


**Figure 2.7:** *Traffic Density Characteristics*

**Example 2 (shock formation):** Let us consider the traffic system given by

$$\rho_t + c(\rho)\rho_x = 0 \tag{82}$$

Where, using Greenshield's model, we have

$$c(\rho) = \frac{\partial q}{\partial \rho} = v_f - 2\frac{v_f}{\rho_{max}}\rho \tag{83}$$

Let the initial condition be given by (Figure 2.5)

$$\rho(x,0) = \begin{cases} 0 & x \le 0 \\ x & 0 < x < 1 \quad -\infty < x < \infty \\ 1 & 1 \le x \end{cases} \tag{84}$$

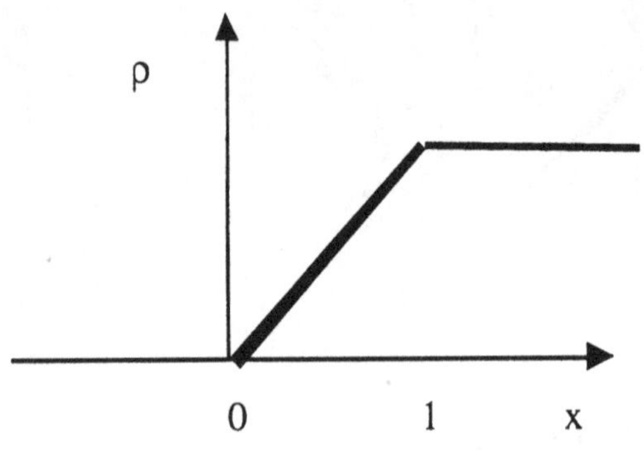


**Figure 2.8:** *Initial Traffic Density*

The slope of the characteristic curve is given by

$$c(\rho_0) = v_f - 2\frac{v_f}{\rho_m}\rho_0 \tag{85}$$

The propagation of the traffic densities for this problem is shown in Figure 2.9.

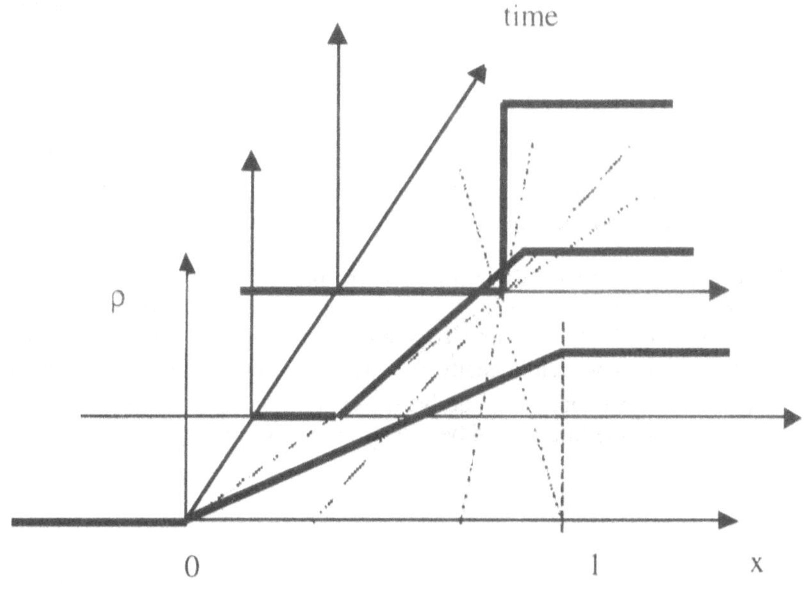


**Figure 2.9:** *Traffic Density vs. Time*

The characteristics of the same problem in the t,x plane is shown in Figure 2.10.

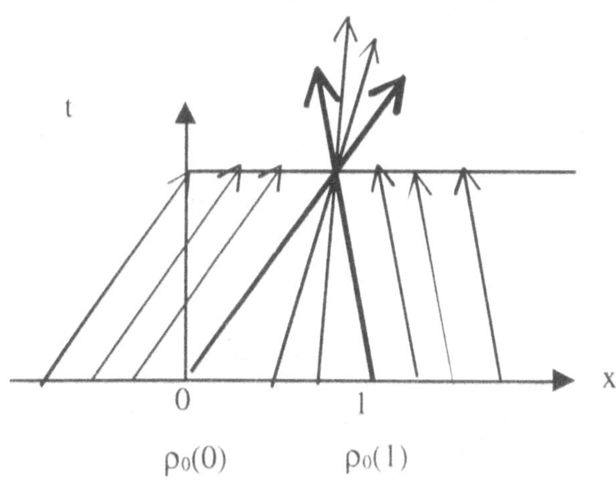


**Figure 2.10:** *Traffic Density Characteristics*

This example shows that a discontinuity (shock) is created after some time when we started with a continuous density profile with positive gradient. We can see that after the shock is formed we have multiple values of density at each x and t (above the horizontal line in Figure 2.10). The question, which needs to be answered is how does the shock propagate after it is formed. This is addressed in the next section.

## 9. Traffic Shock Wave Propagation

We will develop the shock propagation equation n two ways. The first way presents the development in an intuitive way and the second method uses the conservation equation in the integral form to derive the equation.

Shock waves show how discontinuities in the traffic are propagated. To understand this, let us consider the highway section shown in Figure 2.11. In this figure, the highway is divided into two separate homogeneous sections, one with density $\rho_1$ and the other with density $\rho_2$. The line separating the two regions is travelling with the velocity v. To derive the expression for v, we use the fact that the flow going into the line should be equal to the flow coming out from the line.

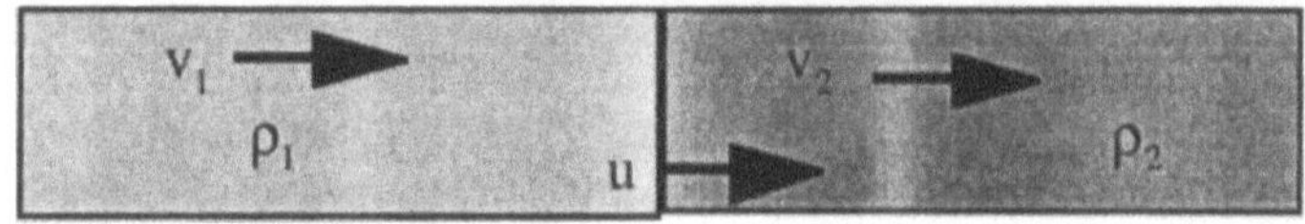


**Figure 2.11:** *Traffic Shock Wave*

The speed of vehicles on the left of the line relative to the line is $v_1 - u$ and for the vehicles on the right is given by $v_2 - u$. Hence, we obtain

$$\rho_1(v_1 - u) = \rho_2(v_2 - u) \tag{85}$$

Using the fact that $\rho_1 v_1 = q_1$ and $\rho_2 v_2 = q_2$, we obtain

$$u = \frac{q_2 - q_1}{\rho_2 - \rho_1} \tag{86}$$

Notice that if we consider a continuous change in flow and density, then we can use

$$q_1 = q \quad q_2 = q + \Delta q \quad \rho_1 = \rho \quad \rho_2 = \rho + \Delta\rho \text{ where } \Delta \to 0 \tag{87}$$

Using (87) in (86) gives

$$u = \frac{\partial q}{\partial \rho} \tag{88}$$

Hence, the slope of the flow-density curve gives the shock wave speed. For the uncongested region, the slope is positive (the shock wave travels forward), and in the uncongested region, the slope is negative giving a backward travelling shock wave. This is shown in Figure 2.12.

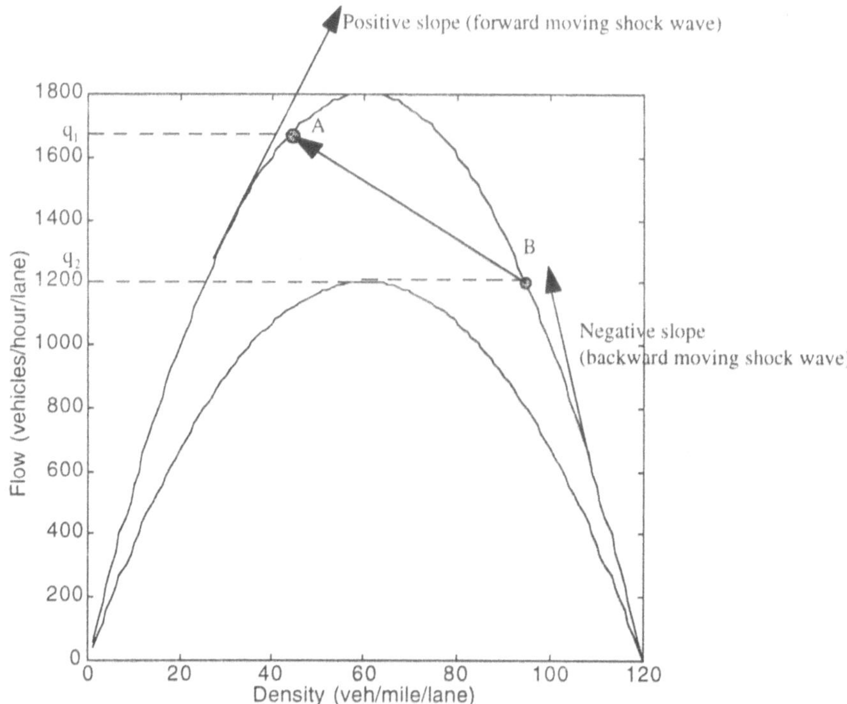


**Figure 2.12:** *Shock Wave Analysis on Flow-Density Curves*

This figure shows two curves: the curve with the higher value of maximum flow is for section 1 of Figure 2.11 and the other curve is for section 2, indicating a sudden drop in the capacity. Since at the bottleneck, $q_1$ is greater than the maximum flow possible in section 2, we achieve $q_2$, the maximum value in the section. The slope from point B to pointA shows a negative value, giving a backward travelling shockwave. Note that we can use various models for speed-density relationships to determine the shock wave speeds.

Now we will present the alternate method to derive (86). We study the traffic propagation along a characteristic defined by a curve x=s(t) in the t-x space. Figure 2-13 shows that the shock is created at time t and then propagates along the dark arrow. To find out at what speed the shock wave travels after it is created, we will consider the shock path curve (which in Figure 2-13 is a straight line but in general can be a curve) as shown in Figure 2-14. We can apply the conservation equation in the integral form knowing that there is a discontinuity at x=s(t) at time t.

$$\frac{d}{dt}\int_a^{s(t)} \rho(x,t)dx + \frac{d}{dt}\int_{s(t)}^{b} \rho(x,t)dx = q(a,t) - q(b,t) \tag{89}$$

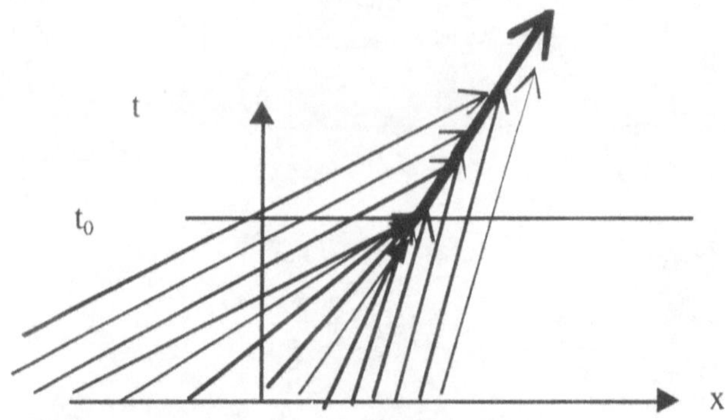


**Figure 2.13:** *Shock Wave Propagation*

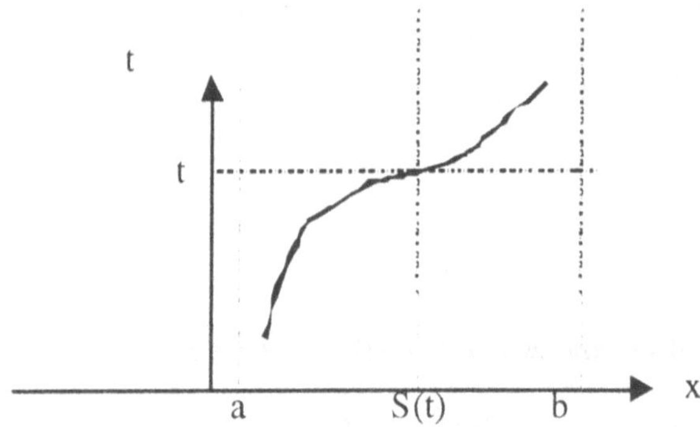


**Figure 2.14:** *Shock Path*

We apply Leibniz' rule to (89) since the limits of integration involve variable t, to obtain (90).

$$\int_{a}^{s(t)} \rho_t(x,t)dx + \int_{s(t)}^{b} \rho_t(x,t)dx + \rho(s^+,t)\frac{ds}{st} - \rho(s^-,t)\frac{ds}{st} \tag{90}$$

$$= q(a,t) - q(b,t)$$

Here s(t) is called the *shock path*, ds/dt is called the *shock speed* and the magnitude of the jump in the traffic density is called the *shock strength*.. Notice that $\rho_t$ is continuous from a to s(t) and from s(t) to b. Hence, if we take the limits $a \to s(t)^-$ and $b \to s(t)^+$, we get

$$\frac{ds}{st} = \frac{q(s^+,t) - q(s^-,t)}{\rho(s^+,t) - \rho(s^-,t)} \tag{91}$$

Which is the same result as (86).

# 10. Traffic Measurements

Traffic variables such as traffic flow, traffic density and traffic speed are needed to use traffic control measures. There are many traffic sensors, loop detectors being traditionally the most commonly used. Loop detectors provide measurements on the spot for a fixed point or section of the road. Some new detectors like the video cameras provide more distributed information. The understanding of the following terms is necessary to be able to design and implement effective controllers.

**Time mean speed**

Time mean speed is the arithmetic mean of the speed of vehicles passing over a point (which could be measured by a loop detector, or some other traffic sensor). The time mean speed is defined as

$$\bar{u}_{time} = \frac{1}{n}\sum_{i=1}^{n} u_i \tag{92}$$

where
n-= number of vehicles passing the detector or a point on the road
$u_i$ = speed of the ith vehicle

**Space mean speed**

Space mean speed is calculated by dividing the total distance traveled by some number of vehicles divided by the total time required by all those vehicles to travel the same distance. This measurement can be done by having two loop detectors kept at a fixed disrtance apart and making measurements on those. The space mean speed is given by

$$\bar{u}_{space} = \frac{nL}{\sum_{i=1}^{n} t_i} = \frac{n}{\sum_{i=1}^{n}(1/u_i)} \tag{93}$$

where
L = length of the section of the road
$t_i$ = time taken by the ith vehicle to traverse the distance L
Space mean speed is used more in traffic engineering and can be derived from time mean speeds (requiring only one detector) using a statistical relationship between the two as [23]

$$\bar{u}_{space} = \bar{u}_{time} - \frac{\sigma^2_{time}}{\bar{u}_{time}} \tag{94}$$

where $\sigma_t^2$ is the variance about the time mean speed.

**Time Headway**

Time headway is given by the difference between the times two consecutive vehicles (leading and the following vehicles) pass the same point (e.g. loop detector) on the road.

**Space Headway**

Space headway is the distance between the front of one vehicle and the front of the following vehicle.

**Flow measurements:**

Traffic flow is measured by counting the number of vehicles N passing a detector in some fixed amount of time T [24]. The flow is given by q=N/T. We can also calculate flow for each vehicle by measuring time headway between the vehicle and its follower and taking its reciprocal. If $h_i$ denotes the time headway for the ith vehicle and $q_i$ the instantaneous flow for ith vehicle, then

$$q_i = 1/h_i \tag{95}$$

For average flow we have

$$q = \frac{N}{T} = \frac{N}{\sum_{i=1}^{N} h_i} = \frac{1}{\frac{1}{N}\sum_{i=1}^{N} h_i} = \frac{1}{\bar{h}} \tag{94}$$

This equation shows that the average flow is the harmonic mean ofthe individual flows.

**Traffic Density Measurements**

Traditionally, traffic density is not measured directly but calculated from

$$\rho = \frac{q}{\bar{u}_{space}} \tag{95}$$

Traffic density can also be calculated using occupancy which is described next.

**Occupancy**

With ITS distributed sensors such as cameras many traffic variables such as traffic density, traffic mean speed, or traffic flow in theory can be obtained. However, it is difficult for loop detectors or spot detectors to do the same. They use indirect method of calculating traffic density by measuring occupancy. Occupancy is defined as the percent of time a detector is sensing a vehicle presence to the total time in some chosen time interval. There is a relationship between occupancy and traffic density, which is used to then calculate traffic flow from occupancy. Traffic density is linearly proportional to traffic density. For instance, if we use number of vehicles per mile as the unit for traffic density then we have,

$$\rho = \gamma \frac{5280}{L_{eff}} \tag{96}$$

where $\gamma$ is the occupancy, $L_{eff}$ is the effective vehicle length in feet.

### Distributed Measurements

Camera sensors give us distributed information which can be used to measure the traffic variables. The density is simply the number of vehicles divided by the length of some section in which the vehicles are counted. We can also calculate density for each vehicle as the reciprocal of the distance of the vehicle from the vehicle ahead. Speed can be calculated from capturing positions of vehicles in different time frames. Flow can be calculated by just multiplying traffic speed and density.

### Moving Observer Method

This method utilizes a moving vehicle taking measurements of variables such as the number of vehicles passing it, and the number of vehicles it passes. This method is not amenable for real-time traffic control in an automated fashion.

## 11. Summary

This chapter presented the traffic theory in a concise fashion which should build the foundation of the reader for designing and analyzing traffic controllers for traffic assignment and routing.

## 12. Exercises

### Questions

Question 1:What is the conservation law for highways? Describe in words.
Question 2:Explain diffusion term (in equation (5)) in a few sentences describing how does a human driver behavior relate to it?
Question 3: What is the difference between time mean speed and space mean speed? What kind of errors can you expect if you use one instead of the other?
Question 4: What is the difference between time headway and space headway. Derive a relationship between the two.
Question 5: How can a computer be used to automatically calculate the traffic variables from distributed measurements?
Question 6 : Prove the implicit function theorem.
Question 7: State Leibniz' rule.
Question 8: Why is the moving observer method not useful for real-time traffic control?

**Problems**

*Problem 1*

Write down traffic flow q in terms of traffic density by utilizing equations (2) and (3). Find out the value of the traffic density at which the traffic flow is maximum.

*Problem 2:*

If we write the Greenshield's model as

$$v_i = v_f(1-\eta_i), \quad \text{where } \eta_i = \frac{\rho_i}{\rho_{max}}$$

then using (6) in (86) show that $u = v_f[1-(\eta_1+\eta_2)]$ Note that this formula can be used to study the following three cases:

(12) **Density Nearly Equal:**
When $\eta_1 \approx \eta_2$, then $u = v_f[1-2\eta_1]$.

(13) **Stopping Waves:**
When a traffic light at an intersection turns red then $\eta_2 = 1$, which gives $u = -v_f\eta_1$.

**(14) Starting Waves**
When a traffic light at an intersection turns green from red, then $\eta_1 = 1$, which gives $u = -v_f\eta_2$. Since $\eta_2 = 1 - \frac{v_2}{v_f}$, we get $u = -v_f + v_2$. Note that velocity $v_2$ is usually small and can be neglected.

*Problem 3:*

One one section of the I-81 corridor in Virginia, there is only one lane available to the traffic which is flowing at a rate of 1200 vehicles/hour at a density of 30 vehicles/mile. A truck enters the highway and travels at a slower speed creating a local traffic density behind the truck of 60 vehicles/mile and a flow of 900 vehicles/hour. Calculate at what rate will the queue length behind the truck increase?

## 13. References


(1) Greenshields, B. D., "A Study in Highway Capacity, Highway Research Board," Proceedings, Vol. 14, 1935, page 458.

(2) Greenberg, H., "An Analysis of Traffic Flow," Operations Research, Vol. 7., 1959, pages 78-85.

(3) Underwood, R. T., "Speed, Volume, and Density Relationships, Quality and Theory of Traffic Flow," Yale Bureau of Highway Traffic, New Haven, Conn., 1961, pages 141-188.

(4) Drake, J.S., Schofer, J. L., and May, A. D., Jr., "A Statistical Analysis of Speed Density Hypothesis," Third Symposium on the Theory of Traffic Flow Proceedings, Elsevier North Holland, Inc., New York, 1967.

(5) Drew, Donald R., Traffic Flow Theory and Control, McGraw Hill Book Company, New York, 1968, Chapter 12.
(6) Pipes, L. A., "Car-Following Models and the Fundamental Diagram of Road Traffic," Transportation Research, Vol. 1 , No.1, 1967, pages 21-29.
(7) May, Adolf D., Traffic Flow Fundamentals, Prentice Hall, New Jersey, 1990.
(8) Musha, T., and Higuchi, H., "Traffic current fluctuation and the Burgers' equation,' Japanese Journal of Applied Physics, 1978, 17, 5, 811-816.
(9) Burns, J. A. and Kang S., 'A control problem for Burgers' equation with bounded input/output', Nonlinear Dynamics 2, 235-262, 1991.
(10) Cole, J. D., 'On a quasi-linear parabolic equation occuring in aerodynamics', Quart. Appl. Math. IX, 1951, 225-236.
(11) Glimm, J. and Lax, P., *Decay of Solutions of Systems of Nonlinear Hyperbolic Conservation Laws*, Amer. Math. Soc. Memoir 101, A.M.S., Providence, 1970.
(12) Hopf, E., 'The partial differential equation $u_t + uu_x = \mu u_{xx}$', Comm. Pure and Appl. Math 3, 1950, 201-230.
(13) Lax, P. D., *Hyperbolic Systems of Conservation Laws and the Mathematical Theory of Shock Waves*, CBMS-NSF Regional Conference Series in Applied Mathematics 11, SIAM, 1973.
(14) Maslov, V. P., 'On a new principle of superposition for optimization problems', *Uspekhi Mat. Nauk* 42, 1987, 39-48.
(15) Maslov, V. P., 'A new approach to generalized solutions of nonlinear systems', *Soviet Math. Dokl* 35, 1987, 29-33.
(16) Curtain, R. F., 'Stability of semilinear evolution equations in Hilbert space', J. Math. Pures et Appl. 63, 1984, 121-128.
(17) Papageorgiou, *Applications of Automatic Control Concepts to Traffic Flow Modeling and Control*, Springer-Verlag, 1983.
(18) Garber, N. J., and Hoel, L. A., Traffic and Highway Engineering, PWS Publishing Company, Boston, MA, 1997.
(19) Gazis, D.C., Herman, R., Potts, R., "Car-following Theory of Steady-State Traffic Flow," Operations Research 7, 1959, pages 499-595.
(20) Farlow, Stanley J., Partial Differential Equations for Scientists and Engineers, Dover Publications, 1992.
(21) Zachmanoglou, E. C., and Thoe, Dale W., Introduction to Partial Differential Equations with Applications, Dover Publications, 1986.
(22) Logan, J. David, An Introduction to Nonlinear Partial Differential Equations, John Wiley and Sons, 1994.
(23) Wadrop, J. G., Some Theoretical Aspects of Road Research, Proc. Inst. Civil Eng., Part II, 1(2):325-362, 1952.
(24) Traffic Flow Theory: A Monograph, Special Report 165, Transportation Research Board, National Research Council, Washington D.C., 1975.
(25) Lighthill, M. J., and Whitham, G. B., 'On kinematic waves II. A theory of traffic flow on long crowded roads', Proc. of the Royal Society of London, Series A 229, 1955, 317-345.

# CHAPTER 3
# MODELING AND PROBLEM FORMULATION

## Objectives

- To provide the system dynamics model for control design in continuous and discrete time and space variables for traffic flow
- Develop the basic framework for the control design

## 1. Introduction

The last chapter presented the basic traffic flow theory based on which we will present modeling and problem formulations for the traffic routing problem. This formulations are for pointwise diversion problems. The network level problems are described in Chapter 8.

## 2. System Dynamics

The first step in the design of feedback controllers for DTR is to model the system dynamics appropriately. A macroscopic model of the traffic can effectively be used in this context. We presented the continuous traffic models in Chapter 2, but we will present the basic model again here for completeness. From the macroscopic perspective, the traffic flow is considered analogous to a fluid flow, which is a distributed parameter system represented by partial differential equations. Mass conservation model of a highway, characterized by $x \in [0, L]$, which is the position on the highway, is given by [1-2]

$$\frac{\partial}{\partial t}\rho(x,t) = -\frac{\partial}{\partial x}q(x,t) \tag{1}$$

where $\rho(x,t)$ is the density of the traffic as a function of x and time t, and $q(x,t)$ is the flow at given x, and t. The flow $q(x,t)$ is a function of $\rho(x,t)$, and the speed v(x,t), as shown below:

$$q(x,t) = \rho(x,t)\,v(x,t) \tag{2}$$

This model of a highway section is shown in Figure 3.1.

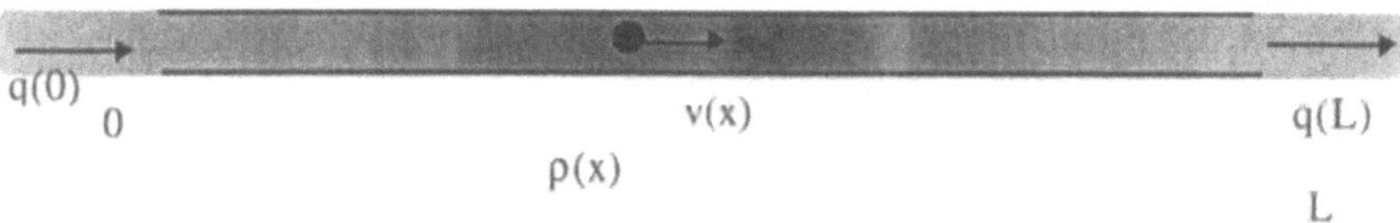


**Figure 3.1:** ***Segment of Highway Model***

There are various static and dynamic models that have been used to represent the relationship between v(x,t) and $\rho(x,t)$. One of the most simple models is the one proposed by Greenshield [3], which hypothesizes a linear relationship between the two variables.

$$v = v_f(1 - \frac{\rho}{\rho_{max}}) \tag{3}$$

where $v_f$ is the free flow speed, and $\rho_{max}$ is the jam density.

## 3. Feedback Control for the Traffic as a Distributed Parameter System

The system represented by equations (1-3) is an infinite dimensional representation of the traffic since it has infinite state variables. There are two ways to design a DTR feedback controller for such a system. One way is to work in the infinite dimensional domain, and design a controller, that can be discretized. Another way is to space discretize PDE (Partial Differential Equation) (1) to obtain an ODE (Ordinary Differential Equation) representation of the system. The ODE representation is a standard representation for most of the results available in feedback control theory, and therefore, using that representation, many techniques from control theory can be utilized.

The modeling of traffic in the PDE domain provides a reasonably accurate model of the traffic system, especially since phenomena such as shock waves are effectively represented. Hence, it would be highly desirable to design a feedback controller directly utilizing this model. However, it is not trivial to design feedback control laws using this modeling scheme. Fortunately, transforming the traffic model into a more convenient form of Burgers' equation reduces the complexity of the control design. Burgers' equation is just a way of representing the same hydrodynamic traffic flow model in a different form utilizing the diffusion behavior of the traffic. Hence, this model is not different in physical interpretation than the classical models that are being widely used in the traffic modeling community.

In the PDE context, researchers have used Burgers' equation to model the traffic flow [4-5]. There has been some research work conducted on the problem of computing feedback laws for Burgers' equation, which is a nonlinear PDE [6, 7]. Burgers' equation

$$\frac{\partial}{\partial t}\rho(x,t) + \rho(x,t)\frac{\partial}{\partial x}\rho(x,t) = \varepsilon\frac{\partial^2}{\partial x^2}\rho(x,t) \tag{4}$$

was introduced by Burgers [8-10] as a simple model for turbulence, where $\varepsilon$ is a viscosity coefficient. The following is borrowed from [4] to show how the traffic problem can be modeled as Burgers' equation.

In order to account for the fact that drivers look ahead and modify their speeds accordingly, (3) can be replaced by

$$v_e = v_f(1 - \frac{\rho}{\rho_{max}}) - D(\frac{\partial \rho}{\partial x})/\rho \tag{5}$$

Using (5) and the fact that $q = \rho v_e$, relationship (2) can be replaced by

$$q(x,t) = \rho(x,t)v(x,t) - D\frac{\partial}{\partial x}\rho(x,t) \tag{6}$$

where D is a diffusion coefficient [10] given by

$$D = \tau v_r^2 \tag{7}$$

where $v_r$ is a random velocity, and $\tau$ is a parameter. Diffusion is a useful concept mentioned by many researchers as an extension to the existing traffic flow models to improve their realism [4,5,11]. The diffusion term represents "the diffusion effect" due to the fact that each driver's gaze is concentrated on the road in front of him/her, so that he/she adjusts his/her speed according to the concentration ahead. This adjustment creates a dependence of flow on concentration gradient that leads to an effective diffusion. This models the gradual rather than instantaneous reduction of speed by the drivers in response to the shock waves. The diffusion term D has the units of velocity multiplied with those of length, such as mile$^2$/hour. Some researchers have also used Burgers' equation to represent the traffic system [4,5]. Combining equations (1) and (6) gives

$$[\frac{\partial}{\partial t}\rho(x,t) + v_f\frac{\partial}{\partial x}\rho(x,t)] - 2\frac{\rho}{\rho_{max}}v_f\frac{\partial}{\partial x}\rho(x,t) - D\frac{\partial^2}{\partial x^2}\rho(x,t) = 0 \tag{8}$$

If we introduce a moving reference frame

$$\xi(x,t) = -x + v_f t \tag{9}$$

and non-dimensionalize $\rho(x,t)$ by $\rho_{max}/2$, and t by $t_0$, equation (7) is transformed to

$$\frac{\partial}{\partial t}\rho(\xi,t) + \rho\frac{\partial}{\partial \xi}\rho(\xi,t) - \frac{1}{R_e}\frac{\partial^2}{\partial \xi^2}\rho(\xi,t) = 0 \tag{10}$$

Here, $R_e$ is a dimensionless constant, and is analogous to the Reynolds number in fluid dynamics. $R_e$ is given by

$$R_e = (\frac{v_f}{v_r})^2\frac{t_0}{\tau} \tag{11}$$

Equation (10) shows Burgers' equation formulation of the traffic flow problem. Some researchers have also worked on the conservation law

$$\frac{\partial}{\partial t}\rho(x,t) + \rho(x,t)\frac{\partial}{\partial x}\rho(x,t) = \varepsilon\frac{\partial^2}{\partial x^2}\rho(x,t) \tag{12}$$

with a solution obtained by taking the following limit:

$$\rho(x,t) = \lim_{\varepsilon \to 0}\rho^{\varepsilon}(x,t) \tag{13}$$

where $\rho^{\varepsilon}(x,t)$ satisfies (4) [12-17]. Using this form reduces the Burgers' equation formulation into the classical traffic model with no diffusion. So, even in the PDE domain, we can try to work on the same model as the classical traffic models.

Work on the feedback control of Burgers' equation has been performed [6,7,17] by some researchers in the past. Curtain [18] showed using Kielhofer's stability results for semi-linear evolution equations [19], that there exists a stabilizing feedback law which can be obtained from the linearized equation, when the domain of the output operator is a certain subspace of $L^2$ which contains the Sobolev space $H_0^1$, where for a given domain $\Omega$ with boundary $\partial\Omega$, $L^2(\Omega)$ is the space of all measurable functions f such that $\int_\Omega |f(x)|^2 dx < \infty$, and $H_0^1(\Omega)$ is the set of all functions f in $L^2(\Omega)$ such that the derivatives $f'$ (or $\nabla f$) are also in $L^2(\Omega)$ and $f|_{\partial\Omega} = 0$, implying that f=0 on the boundary. Burns and Kang [6] show the design of Linear Quadratic Regulator (LQR) optimal controller for the linearized equation [7]. They also study a boundary control problem [7] which is relevant for the traffic control problem, since the control split factor enters the dynamics as a boundary injection.

The DTR problem in the distributed parameter domain is essentially a boundary injection feedback control problem, where the control input enters the traffic system dynamics through the boundary condition. One solution to that relies on converting it into an affine control problem and applying LQ method. Some background development towards that is presented in Chapter 5 of this book.

**DTR Formulation**

One of the most common diversion scenarios involves the point diversion of traffic in order to route vehicles from one entry point of a congested freeway to an alternate route. This is a problem with major practical and application-related implications. This is also relatively easy to implement as the local traffic system can be seen as a de-coupled system, and it has limited infrastructure requirements. This sample problem as shown in Figure 3.2, where there are two alternate routes and the incoming traffic has to be divided between the two, is chosen to illustrate the formulation of the DTR problem in a PDE setting.

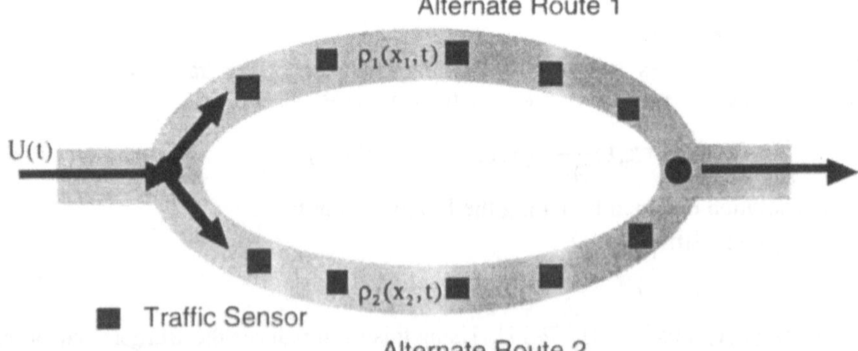


**Figure 3.2:** *Alternate Routes*

We can write two Burgers' equations, one for each route. The following are the conservation equations (from which Burgers' equations are to be derived) for the two routes:

$$\frac{\partial}{\partial t}\rho_1(x_1,t)+\frac{\partial}{\partial x}q_1(x_1,t)=0, \quad 0\le x_1\le L_1, \tag{14}$$

$$\frac{\partial}{\partial t}\rho_2(x_2,t)+\frac{\partial}{\partial x}q_2(x_2,t)=0, \quad 0\le x_2\le L_2, \tag{15}$$

with the following boundary conditions at $(x_1,x_2)=(0,0)$:

$$q_1(0,t)=\beta(t)U(t) \quad \text{and} \quad q_2(0,t)=(1-\beta(t))U(t), 0\le\beta(t)\le 1. \tag{16}$$

These boundary conditions are written in terms of flow. By substituting the relationship (6) in (16), we get the boundary conditions as nonlinear relationships as shown below.

$$\begin{aligned} &\rho_1(x_1,t)\,v(x_1,t)-D\frac{\partial}{\partial x_1}\rho_1(x_1,t)=\beta(t)U(t), \\ &\rho_2(x_2,t)\,v(x_2,t)-D\frac{\partial}{\partial x_2}\rho_2(x_2,t)=(1-\beta(t))U(t). \end{aligned} \tag{17}$$

The boundary conditions at $(x_1,x_2)=(L_1,L_2)$ should be absorbing boundary conditions so that the outflow of the sections is dependent on the dynamics of the traffic rather than on the imposed boundary condition. The output bounded equations for this traffic system, based on the sensors used, are:

$$\begin{aligned} &y_i^1(t)=\wp_i^1(q_1(x_1,t)), \quad i=1,2,....,m, \\ &y_j^2(t)=\wp_j^2(q_2(x_2,t)), \quad j=1,2,...,p, \end{aligned} \tag{18}$$

where the bounded function $\wp_i^1(.)$ is given by

$$\wp_m^k(q_k(x_k,t))=\frac{1}{2\delta}\int_{\bar{x}_m-\varepsilon}^{\bar{x}_m+\varepsilon}q_k(x_k,t)dx_k. \tag{19}$$

The control aim is to solve for user equilibrium, i.e. to equate travel times on the two alternate routes. The nonlinear optimal control problem can be stated as: find $\beta_o(t)$, the optimal $\beta(t)$, which minimizes

$$J(\beta)=\int_0^{t_f}[\int_0^{L_1}\chi(x_1,t)dx_1-\int_0^{L_2}\chi(x_2,t)dx_2]^2dt, \tag{20}$$

where $\chi(.,.)$ is the travel time function and $t_f$ is the final time. Note that a feedback solution is needed for the problem, not an open loop optimal control. Hence, we can either decide the structure of the feedback control such as a PID control with constant gains, and solve numerically for the optimal values of the gains, or we can state the control objective for a standard feedback control problem, such as steady state asymptotic stability

$$\lim_{t\to\infty}[\int_0^{L_1}\chi(x_1,t)dx_1-\int_0^{L_2}\chi(x_2,t)dx_2]\to 0 \tag{21}$$

and some transient behavior characteristics such as some specified settling time, or percent overshoot. We could also linearize the nonlinear system and formulate the LQR problem solution of which is a state feedback control, or try some extensions of techniques developed for control of nonlinear ODEs to nonlinear PDEs.

This problem can be generalized to "n" alternate routes as follows:

Problem 3.1: Find $\beta_o^i$, i=1,2,...,n, which minimize

$$J(\beta_0^i, i=1,2,..n)=\int_0^{t_f}\sum_{k,p}\{\int_0^{L_k}\chi(x_k,t)dx_k - \int_0^{L_p}\chi(x_p,t)dx_p\}^2 dt \tag{22}$$

(k=1,2,...,n, p=1,2,...,n, and the summations are taken over the total number of combinations of n and p, and not permutations so that (k,p)=(1,2) is considered the same as (k,p)=(2,1), and hence only one of these two will be in the summation), or which guarantee

$Lt_{t\to\infty} e \to 0$,

$$\text{where } e=[\{\int_0^{L_1}\chi(x_1,t)dx_1 - \int_0^{L_2}\chi(x_2,t)dx_2\},...,\{\int_0^{L_k}\chi(x_k,t)dx_k - \int_0^{L_p}\chi(x_p,t)dx_p\},...] \tag{23}$$

with some transient behavior characteristics such as some specified settling time or percent overshoot. for the system

$$\frac{\partial}{\partial t}\rho_i(x_i,t)+\frac{\partial}{\partial x}q_i(x_i,t)=0, \quad 0\le x_i \le L_i, \quad i=1,2,...,n, \tag{24}$$

with the boundary conditions

$$\rho_i(x_i,t)v(x_i,t)-D\frac{\partial}{\partial x_i}\rho_i(x_i,t)=\beta^i(t)U(t), \tag{25}$$

absorbing boundary conditions on the other boundary, output equations

$$y_i^j(t)=\wp_i^j(q_i(x_i,t)), \quad i=1,2,...,n, \quad j=1,2,...,k_j, \tag{26}$$

and input constraint

$$\sum_{i=1}^{n}\beta^i=1. \tag{27}$$

The input constraint ensures that the inflow of the traffic is distributed among the candidate alternate routes.

## 4. Discretized System Dynamics

Many researchers have studied and designed optimal open loop controllers utilizing space and time discretized models of traffic flow [20, 21]. Some researchers have also designed feedback control laws using similar models[2, 22]. The reason for the popularity of these models is that there are many techniques available to deal with discretized systems. The same is also true for feedback control, and hence, in order to utilize the various linear and nonlinear [23-26] control techniques available for lumped parameter systems, the distributed parameter model is space discretized [22]. For this, the highway is subdivided into several sections as shown in Figure 2.3.

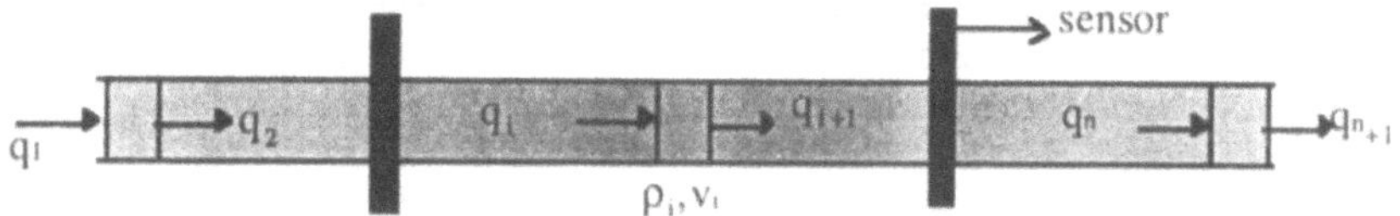


**Figure 2.3:** *Highway Divided into Sections*

Space discretization is performed by dividing the considered highway links into segments. In general, the length of each segment is taken to be between .5 mile and 1 mile. This is an approximation that is quite realistic since the traditional sensors like loop detectors along a freeway are generally installed at least 1 mile apart. Although a smaller step size for space discretization will undoubtedly improve the accuracy of the simulation, in reality it is not possible to measure speed and flow variables at smaller intervals due to limited availability of sensors along freeways. Thus, 1-mile segment length for space discretization appears to be a realistic assumption. On the other hand, the time discretization can be done using very small time steps since traffic data can be downloaded from sensors practically at every second. With the use of more sensors such as CCTVs we can obtain information at higher resolution to have smaller space discretization steps.

The space discretized form of (1) produces the following n continuous ODEs for the n sections of the highway:

$$\frac{d}{dt}\rho_i = \frac{1}{\delta_i}[q_i(t) - q_{i+1}(t) + r_i(t) - s_i(t)], \quad i = 1,2,...n. \tag{28}$$

Here, the $r_i(t)$ and $s_i(t)$ terms indicate the on-ramp and off-ramp flows. Equation (28) combined with (2), (3) (or (4), (5) depending on the decision to include the diffusion term), and the output equations (29) give the mathematical model for a highway, which can be represented in a standard nonlinear state space form for control design purposes:

$$y_j = g_j(\rho_1, \rho_2, ..., \rho_n), \quad j = 1,2,...,p, \tag{29}$$

The standard state space form is

$$\begin{aligned} &\frac{d}{dt}\mathbf{x}(t) = \mathbf{f}[\mathbf{x}(t), \mathbf{u}(t)], \\ &\mathbf{y}(t) = \mathbf{g}[\mathbf{x}(t), \mathbf{u}(t)], \\ &\mathbf{x}(0) = \mathbf{x}_0, \end{aligned} \tag{30}$$

where $\mathbf{x} = [\rho_1, \rho_2, ..., \rho_n]^T$ and $\mathbf{u}(t) = q_0(t)$.

There are various other proposed models that are more detailed in the description of the system dynamics. The phenomenon of shock waves, which is very well represented in the PDE representation of the system, is modeled by expressing the traffic flow between two contiguous sections of the highway as the weighted sum of the traffic flows in those two sections which correspond to the densities in those two sections [1, 27-28]. A dynamic relationship instead of a static one like (3) has also been proposed by [29] and used successfully.

The model thus obtained can also be time discretized to transform the continuous time model into a discrete time mode. A comprehensive model, which incorporates shock waves, as well as represents the dynamic nature of mean speed propagation, is shown in [22] and is reproduced here for completion. The difference equations

$$\rho_j(k+1)=\rho_j(k)+\frac{T}{\delta_j}[q_{j-1}(k)-q_j(k)],$$

$$v_j(k+1)=v_j(k)+\frac{T}{\tau}[v_e(\rho_j(k))-v_j(k)]+\frac{T}{\delta_j}v_j(k)[v_{j-1}(k)-v_j(k)] \quad (31)$$

$$-\frac{vT}{\tau\delta_j}[\frac{\rho_j(k+1)-\rho_j(k)}{\rho_j(k)+\vartheta}]$$

with the relationships

$$q_j(k)=\alpha\rho_j(k)v_j(k)+(1-\alpha)\rho_{j+1}(k)v_{j+1}(k), \quad 0\leq\alpha\leq 1,$$

$$v_e(\rho)=v_f[1-(\frac{\rho}{\rho_{max}})^l]^m, \quad (32)$$

output measurements of traffic flows q and time mean speeds y, shown as

$$y_j(k)=\gamma v_j(k)+(1-\gamma)v_{j+1}(k), \quad 0\leq\gamma\leq 1, \quad (33)$$

and the boundary conditions

$$v_0(k)=y_0(k),$$

$$\rho_{n+1}(k)=q_n(k)/y_n(k), \quad (34)$$

give the discrete system dynamics, which can be represented in the standard nonlinear discrete time form

$$\mathbf{x}(k+1)=\mathbf{f}(\mathbf{x}(k),\mathbf{u}(k)), \quad (35)$$

$$\mathbf{y}(k)=\mathbf{g}(\mathbf{x}(k),\mathbf{u}(k)),$$

$$\mathbf{x}(0)=\mathbf{x}_0,$$

where control u(k) for a single highway can be the input flow, and mean speeds at the entrance and the exit of the highway.

If the control actuation is discrete, such as these implemented by microprocessors and computers, feedback control laws can be designed based on the discrete model (35), or can be designed using (30), after which the controller can be discretized.

## 5. Feedback Control for the Traffic as a Lumped Parameter System

In the discretized traffic flow model, the freeway is divided into sections with aggregate traffic densities. Sensors are used to measure variables such as densities, traffic flow, and traffic average speeds in these sections, which can be used by the feedback controller to give appropriate commands to actuators such as VMS, and HAR.

There are essentially two ways to design controllers for such nonlinear traffic systems of the form (30) and (35). One way is to design the controller by linearizing the nonlinear dynamics of the system about its equilibrium or a trajectory; the other way is to design directly for the nonlinear system. The first way is easier since immense literature is available describing the various design techniques, especially for LTI systems, but the results are valid only where the linearization is applicable. On the other hand, design of controllers directly for the nonlinear system is much more difficult, but the results are usually global. Some of the linear control techniques are LQR, LQG, PID, $H_\infty$, and preview control, and some of the nonlinear ones are describing functions design, feedback linearization, sliding mode control, and nonlinear $H_\infty$. Qualitative methods such as fuzzy control and expert systems can also be utilized. Qualitative methods are usually easy to design but difficult to tune and analyze. Some of these issues are discussed in [2] by Messmer and Papageorgiou.

**DTR Formulation**

For the two alternate routes problem, as shown in Figure 3.2, the two routes are divided into $n_1$ and $n_2$ sections, respectively. For simplicity, we are considering the static velocity relationship and ignoring the effect of downstream flow for the model

$$\frac{d}{dt}\rho_{i,j}=\frac{1}{\delta_i}[q_{i,j-1}(t)-q_{i,j}(t)],\quad (i,j)=((1,1),(1,2),\ldots,(1,n_1),(2,1),(2,2),\ldots,(2,n_2)) \tag{36}$$

with relationships (2) and (3). The control input is given by

$$\begin{aligned}&\beta(t)U(t)=q_{1,0}(t),\quad 0\le\beta\le 1,\\&(1-\beta(t))U(t)=q_{2,0}(t)\end{aligned} \tag{37}$$

The flow U(t) is measured as a function of time, and the splitting rate $\beta(t)$ is the control input. The output measurement could be the full state vector, i.e. vector of flows of all the sections, or a subset of that. The control problem can be stated as: find $\beta_o(t)$, the optimal $\beta(t)$, which minimizes

$$J(\beta)=\int_0^{t_f}[\sum_{i=1}^{m}\chi(\rho_i)-\sum_{m+1}^{m+p}\chi(\rho_j)]^2dt \tag{38}$$

where $\chi(.,.)$ is the travel time function and $t_f$ is the final time. Note that, just like for the distributed parameter system case, a feedback solution is needed for the problem, and not an open loop optimal control. Hence, here also we can either decide the structure of the feedback control, such as a PID control with constant gains, and solve numerically for the optimal values of the gains, or we can state the control objective for a standard feedback control problem, such as steady state asymptotic stability

$$\lim_{t\to\infty}[\sum_{i=1}^{m}\chi(\rho_i)-\sum_{m+1}^{m+p}\chi(\rho_j)]\to 0 \tag{39}$$

and some transient behavior characteristics such as some specified settling time or percent overshoot.

The discrete time versions of (36-38) are given below.

$$\rho_{i,j}(k+1)=\rho_{i,j}(k)+\frac{T}{\delta_i}[q_{i,j-1}(t)-q_{i,j}(t)],\quad (i,j)=((1,1),...,(1,n_1),(2,1),...,(2,n_2)) \tag{40}$$

$$\begin{aligned}&\beta(k)U(k)=q_{1,0}(k),\quad 0\le\beta\le1,\\&(1-\beta(k))U(k)=q_{2,0}(k),\end{aligned} \tag{41}$$

$$J(\beta)=\sum_{k=1}^{k_f}[\sum_{i=1}^{m}\chi(\rho_i)-\sum_{m+1}^{m+p}\chi(\rho_j)]^2, \tag{42}$$

$$\lim_{k_f\to\infty}[\sum_{i=1}^{m}\chi(\rho_i)-\sum_{m+1}^{m+p}\chi(\rho_j)]\to 0. \tag{43}$$

For an n alternate route problem, the continuous time formulation is:

Problem 5.1a: Find $\beta_o^i$, i=1,2,...,n, which minimize

$$J(\beta_0^i, i=1,2,..n)=\int_0^{t_f}[\sum_{k,p}\{\sum_{i=1}^{n_k}\chi(\rho_{i,k})-\sum_{j=1}^{n_p}\chi(\rho_{j,p})\}^2dt \tag{44}$$

(k=1,2,...,n, p=1,2,...,n, and the summations are taken over the total number of combinations of n and p, and not permutations so that (k,p)=(1,2) is considered the same as (k,p)=(2,1), and hence only one of these two will be in the summation) ), or which guarantee

$$Lt_{t\to\infty}\mathbf{e}\to\mathbf{0},$$

$$\text{where } \mathbf{e}=[\{\sum_{i=1}^{n_1}\chi(\rho_{i,1})-\sum_{j=1}^{n_2}\chi(\rho_{j,2})\},...,\{\sum_{i=1}^{n_k}\chi(\rho_{i,k})-\sum_{j=1}^{n_p}\chi(\rho_{j,p})\},...] \tag{45}$$

with some transient behavior characteristics such as some specified settling time or percent overshoot for the system

$$\frac{d}{dt}\rho_{i,j}=\frac{1}{\delta_i}[q_{i,j-1}(t)-q_{i,j}(t)],\quad (i,j)=((1,1),...,(1,n_1),(2,1),...,(2,n_2),...,(n,1),...,(n,n_n)) \tag{46}$$

with given full and partial state observation, and input constraints

$$\sum_{i=1}^{n}q_{i,0}(t)=U(t) \text{ and } \sum_{i=1}^{n}\beta^i=1. \tag{47}$$

The corresponding discrete time formulation is:

Problem 5.1b: Find $\beta_o^i$, i=1,2,...,n, which minimize

$$J(\beta_0^i, i=1,2,..n)=\sum_{k=1}^{k_f}\sum_{k,p}\{\sum_{i=1}^{n_k}\chi(\rho_{i,k})-\sum_{j=1}^{n_p}\chi(\rho_{j,p})\}^2 \tag{48}$$

(k=1,2,...,n, p=1,2,...,n, and the summations are taken over the total number of combinations of n and p, or which guarantee

$$Lt_{k_f \to \infty} \mathbf{e} \to \mathbf{0},$$

$$\text{where } \mathbf{e} = [\{\sum_{i=1}^{n_1} \chi(\rho_{i,1}) - \sum_{j=1}^{n_2} \chi(\rho_{j,2})\}, \ldots, \{\sum_{i=1}^{n_k} \chi(\rho_{i,k}) - \sum_{j=1}^{n_p} \chi(\rho_{j,p})\}, \ldots] \tag{49}$$

with some transient behavior characteristics such as some specified settling time or percent overshoot for the system

$$\rho_{i,j}(k+1) = \rho_{i,j}(k) + \frac{T}{\delta_i}[q_{i,j-1}(k) - q_{i,j}(k)], \quad (i,j) = ((1,n_1),(2,n_2) \ldots,(n,n_n)) \tag{50}$$

with given full and partial state observation, and input constraints (47) after replacing t with k.

Based on these standard forms, the appropriate feedback control laws can be designed. The exact methodology of the design will be studied in subsequent papers, but the next section elucidates an example to show how the feedback control could be used.

# 6. Sample Problem for Space Discretized Dynamics

In order to illustrate the ideas discussed above, we have designed a feedback control law for a space discretized system such as (36) with two alternate routes. The control is based on the feedback linearization technique for nonlinear systems. The technique is based on defining a diffeomorphism and performing the transformation on the state variables in order to convert them into the canonical form. If the relative degree of the system is less than the system order, then the internal dynamics are studied to ensure that it is stable. The details of this technique are given in [25,26,30]. In this problem, the system order is two and the relative degree is one.

The space discretized flow equations used for the two alternate routes are:

$$\dot{\rho}_1 = -\frac{1}{\delta_1}\left[ v_{f1}\rho_1(1 - \frac{\rho_1}{\rho_{m1}}) - \beta u \right], \tag{51}$$

$$\dot{\rho}_2 = -\frac{1}{\delta_2}\left[ v_{f2}\rho_2(1 - \frac{\rho_2}{\rho_{m2}}) + \beta u - u \right], \tag{52}$$

We have considered a simple first order travel time function, which is obtained by dividing the length of a section by the average velocity of vehicles on it. Note that when there are no cars present, the free-flow velocity will be used by the formula. In that case, the traffic density is zero, and therefore the traffic flow is also zero. According to that, we have

$$\chi_1(k) = d_1 / [v_{f1}(1 - \frac{\rho_1}{\rho_{m1}})], \tag{53}$$

$$\chi_2(k) = d_2 / [v_{f2}(1 - \frac{\rho_2}{\rho_{m2}})], \tag{54}$$

where $d_1$ and $d_2$ are section lengths, $v_{f1}$ and $v_{f2}$ are the free flow speeds of each section, and $\rho_{m1}$ and $\rho_{m2}$ are the maximum (jam) densities of each section. Since we need to equate the travel times, we take the new transformed state variable y as the difference in travel times. Differentiating the equation representing y in terms of the state variables introduces the input split factor into the dynamic equation. Therefore, that transformed equation can be used to design the input that cancels the nonlinearities of the system and introduce a design input $v$, which can be used to place the poles of the error equation for asymptotic stability. These steps are shown below:

The variable y is equal to the difference in the travel time on the two sections:

$$y = \frac{k_1}{(k_2 - \rho_1)} - \frac{k_3}{(k_4 - \rho_2)} \tag{55}$$

This equation can be differentiated with respect to time to give the travel time difference dynamics:

$$\dot{y} = \frac{k_1 \dot{\rho}_1}{(k_2 - \rho_1)^2} - \frac{k_3 \dot{\rho}_2}{(k_4 - \rho_2)^2} \tag{56}$$

By substituting (51) and (52) in (57) we obtain

$$\dot{y} = -\frac{k_1\left( v_{f1}\rho_1(1 - \frac{\rho_1}{\rho_{m1}}) - \beta u \right)}{\delta_1 (k_2 - \rho_1)^2} + \frac{k_3\left( v_{f2}\rho_2(1 - \frac{\rho_2}{\rho_{m2}}) + \beta u - u \right)}{\delta_2 (k_4 - \rho_2)^2} \tag{57}$$

This equation can be rewritten in the following form:

$$\dot{y} = F + G\beta \tag{58}$$

where

$$F = \left[ -\frac{k_1 v_{f1} \rho_1}{\delta_1 (k_2 - \rho_1)^2}(1 - \frac{\rho_1}{\rho_{m1}}) + \frac{k_3}{\delta_2 (k_4 - \rho_2)^2}\left( (1 - \frac{\rho_2}{\rho_{m2}}) v_{f2}\rho_2 - u \right) \right] \tag{59}$$

$$G = \left( \frac{k_1}{\delta_1 (k_2 - \rho_1)^2} + \frac{k_3}{\delta_2 (k_4 - \rho_2)^2} \right) u \tag{60}$$

Hence, a feedback linearization control law can be designed to cancel the nonlinearities and provide the desired error dynamics. The law used is

$$\beta = G^{-1}(-F + v) \tag{61}$$

which gives the closed loop dynamics as

$$\dot{y} = v \tag{62}$$

As was mentioned earlier, since the relative degree of the system is one, and the system order is two, we need to test the stability or boundedness of the second transformed state variable. That variable is taken to be (see [25,26,31] for details of the steps of the design of feedback linearization control laws)

$$\eta = \delta_1 \rho_1 + \delta_2 \rho_2 \tag{63}$$

The state variable $\eta$ is bounded since the densities on the sections can not exceed the corresponding jam densities

$$\eta \le \delta_1 \rho_{m1} + \delta_2 \rho_{m2} \tag{64}$$

and hence the overall system is exponentially stable $(y \to 0)$ if we choose $v = -Ky, \quad K > 0$, and y asymptotically goes to zero as $y(t) = y(0)e^{-Kt}$. This implies that when a splitting value based on (61) is utilized, the difference in travel time of two alternate routes will go to zero at an exponential rate. Hence, the closed loop traffic system controlled by the proposed feedback linearization law is exponentially stable and has desired transient behavior.

# 7. Sample Problem for Space and Time Discretized Dynamics: Three Alternate Routes Case Description

In this section, we show a design of a feedback control system for a simple network consisting of three alternate routes modeled as space and time discretized systems (40). Similar feedback controllers can be designed for larger and more complex traffic networks. For simplicity, we are assuming that each of the alternate routes is just a single discrete section, and also for simplicity we are ignoring the diffusion term in the model.

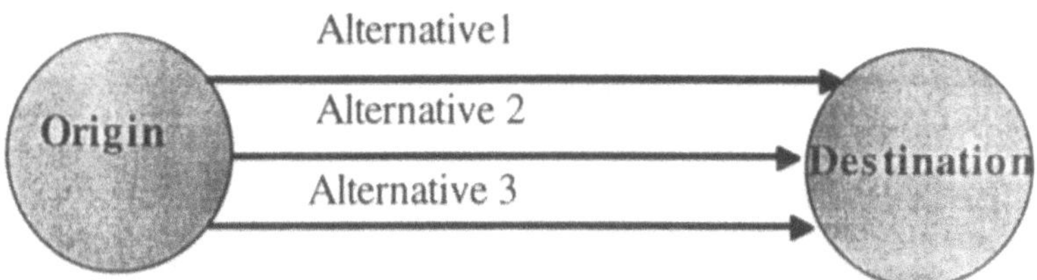


**Figure 2.4:** *Test Network*

**System Dynamics**

The flow equations used for the three alternate routes are:

$$\begin{aligned}
\rho_1(k+1) &= \rho_1(k) - \frac{T}{\delta_1}[q_1^{out}(k) - q_1^{in}(k)], \\
\rho_2(k+1) &= \rho_2(k) - \frac{T}{\delta_2}[q_2^{out}(k) - q_2^{in}(k)], \\
\rho_3(k+1) &= \rho_3(k) - \frac{T}{\delta_3}[q_3^{out}(k) - q_3^{in}(k)],
\end{aligned} \tag{65}$$

where the flows are taken to be

$$q_1^{out}(k) = \rho_1(k)v_1(k),$$
$$q_2^{out}(k) = \rho_2(k)v_2(k), \quad (66)$$
$$q_3^{out}(k) = \rho_3(k)v_3(k),$$

the average velocities are given by

$$v_1 = v_{f1}(1 - \frac{\rho_1}{\rho_{max1}}),$$
$$v_2 = v_{f2}(1 - \frac{\rho_2}{\rho_{max2}}), \quad (67)$$
$$v_3 = v_{f3}(1 - \frac{\rho_3}{\rho_{max3}}),$$

and

$$q_1^{in}(k) = \beta_1(k)U(k), \qquad 0 \le \beta_1(k) \le 1,$$
$$q_2^{in}(k) = \beta_2(k)U(k), \qquad 0 \le \beta_2(k) \le 1, \quad (68)$$
$$q_3^{in}(k) = [1 - \beta_1(k) - \beta_2(k)]U(k)$$

The travel times are given by

$$\chi_1(k) = d_1 / [v_{f1}(1 - \frac{\rho_1}{\rho_{max1}})],$$
$$\chi_2(k) = d_2 / [v_{f2}(1 - \frac{\rho_2}{\rho_{max2}})], \quad (69)$$
$$\chi_3(k) = d_3 / [v_{f3}(1 - \frac{\rho_3}{\rho_{max3}})].$$

The state variables are the $\rho$'s for each route. This is a full state measurement problem, assuming the flows are measured, and can be converted into state variable values using the deterministic relationships. The control variables are the splitting rates. For this example, full compliance is assumed. The overall system in a standard nonlinear state-space form can be written as

$$\mathbf{x}(k+1) = \mathbf{f}(\mathbf{x}(k)) + \mathbf{g}(\mathbf{x}(k))\mathbf{u}(k),$$
$$\mathbf{y}(k) = \mathbf{h}(\mathbf{x}(k)) \quad (70)$$

where

$\mathbf{x}(k) = [\rho_1(k), \rho_2(k), \rho_3(k)]^T$, $\mathbf{y}(k) = [\chi_1(k), \chi_2(k), \chi_3(k)]^T$, $\mathbf{u}(k) = [\beta_1, \beta_2]^T$,

$$\mathbf{f}(\mathbf{x}(k)) = \begin{bmatrix} \rho_1(k) - T\rho_1(k)v_{f1}(1 - \rho_1(k)/\rho_{max1})/\delta_1 \\ \rho_2(k) - T\rho_2(k)v_{f2}(1 - \rho_2(k)/\rho_{max2})/\delta_2 \\ \rho_3(k) - T\rho_3(k)v_{f3}(1 - \rho_3(k)/\rho_{max3})/\delta_3 + U(k) \end{bmatrix}, \quad (71)$$

$$g(x(k)) = \begin{bmatrix} TU(k)/\delta_1(k) & 0 \\ 0 & TU(k)/\delta_2(k) \\ -TU(k)/\delta_1(k) & -TU(k)/\delta_2(k) \end{bmatrix}, \quad h(x(k)) = \begin{bmatrix} d_1/[v_{f1}(1-\frac{\rho_1(k)}{\rho_{max1}})] \\ d_2/[v_{f2}(1-\frac{\rho_2(k)}{\rho_{max2}})] \\ d_3/[v_{f3}(1-\frac{\rho_3(k)}{\rho_{max3}})] \end{bmatrix}$$

Here the system is in the local coordinates for the smooth state space manifold M, f is the smooth drift vector field on M, g is the smooth input vector field on M, and h is the smooth output vector field on M.

### System Analysis

The nonlinear system (65) can be analyzed for controllability and observability. For continuous nonlinear systems, controllability and observability can be studied using Lie brackets and Lie derivatives in terms of accessibility and reachability. These techniques can be extended to be used for a discrete system (65). For review of these topics, please see references [25,26, 30]. We might conjecture that the system is accessible and reachable.

### Simple Feedback Control Law

The controller for (65) is a PI (Proportional-Integral) controller with constant feedforward term. As was discussed in section 5, there are two ways to design a feedback controller for this problem. One way is to solve the optimality and use the feedback solution if available for that problem, and the other is to design a feedback controller to satisfy some steady state asymptotic stability or transient behavior. This controller is designed to produce asymptotic stability to the system so that the error terms defined next tend to zero in time. Since many controllers including LQ are implemented as constant gain (static) controllers, we chose to highlight a design with a constant gain PI controller. The error terms to drive the controller are defined as

$$\begin{aligned} e_1(k) &= \chi_3(k) - \chi_1(k), \\ e_2(k) &= \chi_3(k) - \chi_2(k). \end{aligned} \tag{72}$$

The integral term in the PI control is taken as error summation so that

$$\begin{aligned} ie_1(k) &= \sum_{p=1}^{k} \{\chi_3(p) - \chi_1(p)\}, \\ ie_2(k) &= \sum_{p=1}^{k} \{\chi_3(p) - \chi_2(p)\}. \end{aligned} \tag{73}$$

The PI controller with constant feedforward shown below has been designed to equilibrate travel times on alternate routes. While integral part of the PI controller affects the steady-state behavior of the system, the proportional part of it combined with the integral part affects the transient behavior. This controller is presented here to show the feasibility of feedback control to meet the objective of achieving equal travel time in all the alternate routes.

$$\beta_1(k) = \max[0, \min\{1, (1/3 + k_1 e_1 + k_{i1} ie_1)\}], \tag{74}$$

$$\beta_2(k) = \max[0, \min\{1 - \beta_1(k), (1/3 + k_2 e_2 + k_{i2} ie_2)\}],$$

where $k_1, k_{i1}, k_2, \text{and}\, k_{i2}$ are the controller gains.

As can be seen in the section below, the results of using this controller are encouraging. However, using PI regulators with input saturation suffers from wind-up phenomenon, which could be countered by using the following form for the control laws:

$$\beta(k) = \beta(k-1) - k_1 e_1(k) - k_2[e_1(k) - e_1(k-1)] \tag{75}$$

where $\beta(k)$ is limited in the range [0,1] and $\beta(k-1)$ takes its actually implemented value (e.g. 0 or 1 if it has been on the bounds).

The PI controller described above is shown to be stable using simulations. The results are presented and discussed in the next section.

**Description of the Results for Different Scenarios**

For the simple example problem and its feedback control solution, we have performed several test runs for different scenarios. The simulation program is developed in Matlab. We have tested the simulation using three scenarios. For each scenario, we have considered a simple network that consists of three alternate routes. The splitting decisions are made at one decision point only. Alternate routes have different free flow travel times. The one with the lowest travel time can be assumed to be the freeway and two others with higher travel times can be considered to be highways with a lower level of service.

A brief description of each scenario and the corresponding plots are shown below. Each simulation shows four plots showing travel time, traffic density, splits, and traffic outflow versus time for all three routes. The three routes are shown with three different line styles: solid, dashed, and dot-dashed.

*First Scenario*

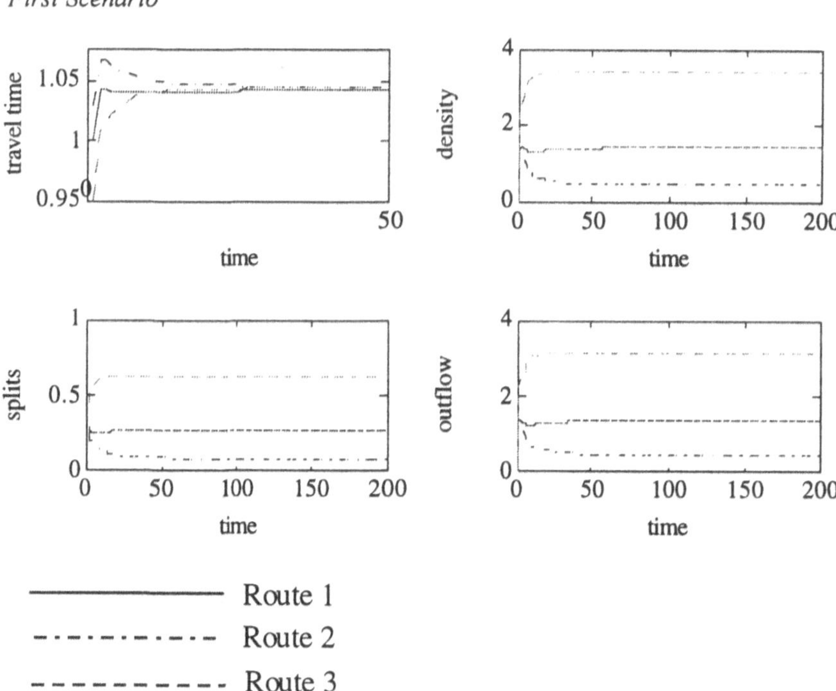


**Figure 3.5:** *Scenario 1*

In this scenario, we have constant inflow and no congestion for all the alternate routes. This scenario represents normal traffic conditions. As can be seen in the first plot, travel times become equal and stay equal until the end of the simulation period since there are no external disturbances.

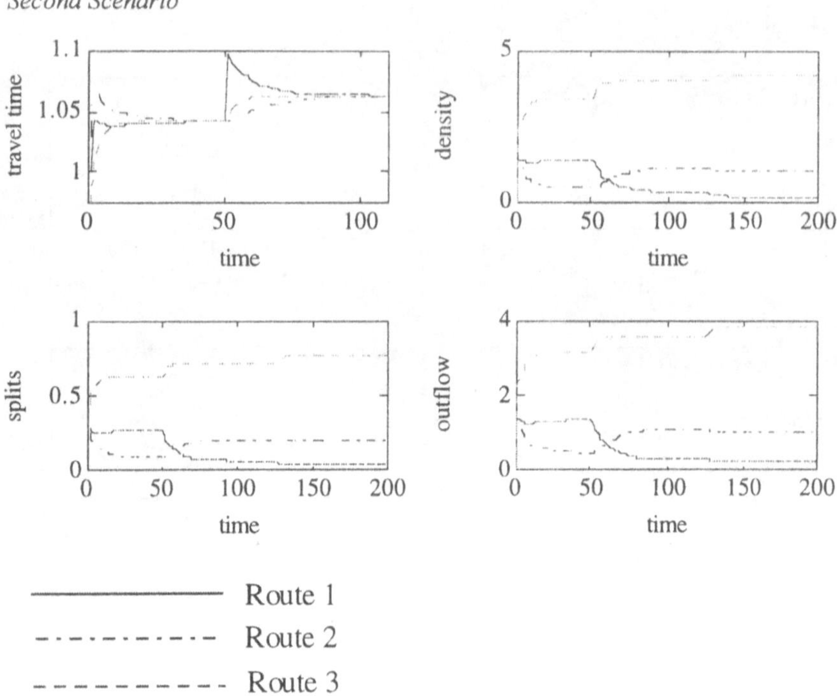


**Figure 3.6:** *Scenario 2*

This scenario simulates congestion on route 2 from time 50 to 75, and relief on route 3 from time 125 to 150. The congestion on route 2 may be due to a temporal bottleneck caused by an incident, and the relief on route 3 may be due to the clearance of an incident that existed before. Fluctuations in travel times and the response of the controller in order to stabilize the system can easily be seen by studying the plots.

*Scenario 3*

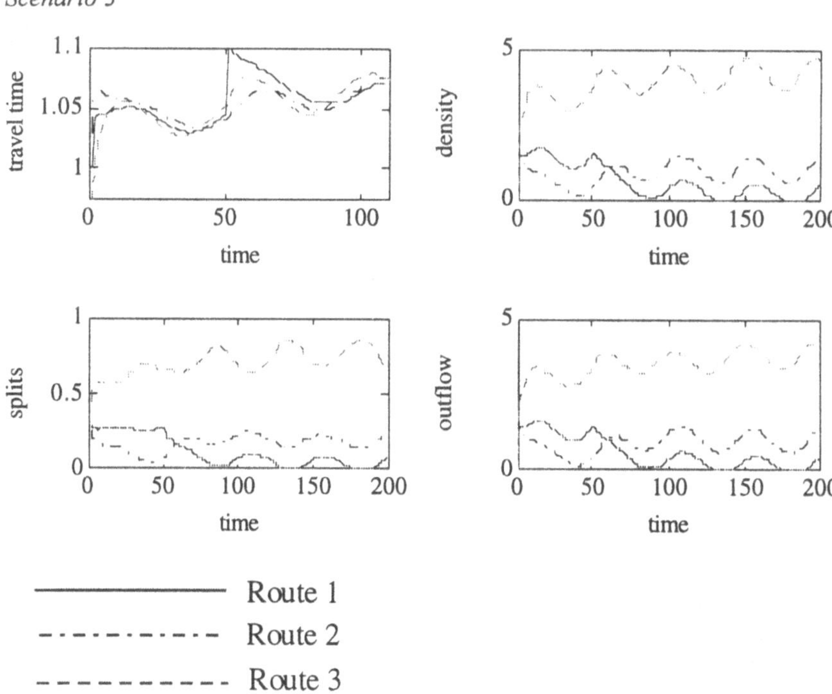


**Figure 3.7:** *Scenario 3*

This last scenario has a sinusoidal inflow (demand) function and the same traffic patterns as scenario 2. The fluctuations in inflow traffic modeled by the sinusoidal are meant to represent the natural traffic fluctuations that occur in reality. In a fluctuation traffic scenario simulation, the controller successfully achieves the objective of equating the travel times on the three alternate routes.

# 8. Summary

- In this chapter, we formulated feedback control problems in the distributed parameter setting as well as in continuous-time and discrete-time lumped parameter settings for DTR.
- Brief discussions on applicable real-time feedback control methods were provided.
- Since most of the routing decisions are local and not network-wide, the solution for the type of DTR problem discussed in this chapter can be promising under incident conditions that do not require network-wide approaches.
- The models presented in this chapter can be extended to the case where on-ramps and off-ramps are also present. More research is needed for developing frameworks for network-wide problems and providing solutions for those. The

pointwise diversion problems are smaller in size (in terms of the size of state variable vector) as compared to network-wide problems. Hence , the calculations for control actions take much less time in the point diversion problem.

## 9. Exercises

### Questions

Question 1: What are the three ways a highway system can be modeled macroscopically?
Question 2:What are discretization errors and how are they introduced when using different types of model for highways?
Question 3:What are the two types of discretization errors in highway modeling problems?

### Problems

*Problem 1:*

Redo the DTR formulation in section 3 when the diffusion term D=0.

*Problem 2:*

Redo section 6 for three alternate routes and section 7 for two alternate routes.

## 10. References

(1) Lighthill, M. J., and Whitham, G. B., 'On kinematic waves II. A theory of traffic flow on long crowded roads', Proc. of the Royal Society of London, Series A 229, 1955, 317-345.
(2) Papageorgiou, M., and Messmer, A., 'Dynamic Network Traffic Assignment and Route Guidance via Feedback Regulation', Transportation Research Board, Washington, D.C., Jan. 1991.
(3) Greenshields, B. D., "A Study in Highway Capacity, Highway Research Board," Proceedings, Vol. 14, 1935, page 458.
(4) Fletcher, C. A. J., 'Burgers' equation: a model for all reasons', in J. Noye (ed.), *Numerical Solution of Partial Differential Equations*, North-Holland, 1982, 139-225.
(5) Musha, T., and Higuchi, H., "Traffic current fluctuation and the Burgers' equation,' Japanese Journal of Applied Physics, 1978, 17, 5, 811-816.
(6) Burns, J. A. and Kang S., 'A control problem for Burgers' equation with bounded input/output', Nonlinear Dynamics 2, 235-262, 1991.
(7) Burns, J. A. and Kang S., ' A stabilization problem for Burgers' equation with unbounded control and observation', Intl. Series of Num. Math., 100, 51-72, 1991.

(8) Burgers, J. M., 'Mathematical examples illustrating relations occuring in the theory of turbulent fluid motion', Trans. Roy. Neth Acad. Sci. 17, Amsterdam, 1939, 1-53.
(9) Burgers, J. M., 'Mathematical model illustrating the theory of turbulence', Adv. in Appl. Mech. 1, 1948, 171-199.
(10) Burgers, J. M., 'Statistical problems connected with asymptotic solution of one-dimensional nonlinear diffusion equation', in M. Rosenblatt and C. van Atta (eds.), *Statistical Models and Turbulence*, Springer, Berlin, 1972, 41-60.
(11) Papageorgiou, Markos, Blosseville, Jean-Mark, and Hadj-Salem Habib "Macroscopic Modelling of Traffic Flow on the Boulevard Peripherique in Paris", Trans. Res. Board, Vol. 23B, No. 1, pp. 29-47, 1989.
(12) Cole, J. D., 'On a quasi-linear parabolic equation occuring in aerodynamics', Quart. Appl. Math. IX, 1951, 225-236.
(13) Glimm, J. and Lax, P., *Decay of Solutions of Systems of Nonlinear Hyperbolic Conservation Laws*, Amer. Math. Soc. Memoir 101, A.M.S., Providence, 1970.
(14) Hopf, E., 'The partial differential equation $u_t + uu_x = \mu u_{xx}$', Comm. Pure and Appl. Math 3, 1950, 201-230.
(15) Lax, P. D., *Hyperbolic Systems of Conservation Laws and the Mathematical Theory of Shock Waves*, CBMS-NSF Regional Conference Series in Applied Mathematics 11, SIAM, 1973.
(16) Maslov, V. P., 'On a new principle of superposition for optimization problems', *Uspekhi Mat. Nauk* 42, 1987, 39-48.
(17) Maslov, V. P., 'A new approach to generalized solutions of nonlinear systems', *Soviet Math. Dokl* 35, 1987, 29-33.
(18) Curtain, R. F., 'Stability of semilinear evolution equations in Hilbert space', J. Math. Pures et Appl. 63, 1984, 121-128.
(19) Kielhofer, H., 'Stability and semilinear evolution equations in Hilbert space', Arch. Rat. Mech. Anal. 57, 1974, 150-165.
(20) Tan, B., Boyce, D. E., and LeBlanc, L. J., " A New Class of Instantaneous Dynamics User-Optimal Traffic Assignment Models", Operations Research, 41, 192-202, 1993.
(21) Friesz, T. L., Luque, J., Tobin, R. L., and Wie, B. W., " Dynamic Network Traffic Assignment Considered as a Continuous Time Optimal Control Problem", Operations Research, 37, 893-901, 1989.
(22) Papageorgiou, *Applications of Automatic Control Concepts to Traffic Flow Modeling and Control*, Springer-Verlag, 1983.
(23) Kuo, B. C., *Automatic Control Systems*, Prentice Hall, Inc., 1987.
(24) Mosca, E., *Optimal Predictive and Adaptive Control*, Prentice Hall, 1995.
(25) Isidori, A., *Nonlinear Control Systems*, Springer-Verlag, 1989.
(26) Slotine, J. J. E., and Li, Weiping, *Applied Nonlinear Control*, Prentice Hall, 1991.
(27) Richards, P. I., Shock waves on the highway. Oper. Res., 1956, 42-51.
(28) Berger, C.R., and Shaw, L., Discrete event simulation of freeway traffic. Simulation council Proc., Series 7, 1977, 85-93.
(29) Payne, H. J., Models of freeway traffic and control. Simulation council proc., 1971, 51-61.
(30) Nijmeijer, H., and Schaft, A. van der, *Nonlinear Dynamical Control Systems*, Springer-Verlag,1990.

# CHAPTER 4
# DYNAMIC TRAFFIC ROUTING PROBLEM IN DISTRIBUTED PARAMETER SETTING

## Objectives

- To design the control law for the DTR (Dynamic Traffic Routing) problem modeled in the distributed parameter setting
- Understand how the discretization errors are related for software simulation
- Develop a software simulation code and perform simulation to study this problem

## 1. Introduction

The aim of the Dynamic Traffic Routing (DTR) solution is to come up with the appropriate values of the traffic split factors at the node where the traffic can take the alternate routes between node A to node B. Then the traffic can be split at the node using VMS, highway advisory radio, or in-vehicle routing systems. The feedback controller is designed to use the current measurements of the traffic conditions on the alternate routes and calculate the split factor values. This strategy can be used for real on-line traffic control, but also can be used for traffic simulations and for planning purposes.

The highway can be modeled using the hydrodynamic theory comparing the traffic flow with fluid flow. That gives us the system dynamics of the traffic control problem in the distributed parameter setting in terms of Partial Differential Equations (PDEs). There are two ways, we can perform control design and analysis in this setting. One method is presented in this chapter which converts the PDE model into an ode using a distributed control. The other method involves highly mathematical treatment of PDEs by using semi-group theory and evolution equations. That method is introduced in the next chapter.

## 2. System Dynamics

In order to design the controller, we need to choose an appropriate model for the system. We use the fluid flow distributed parameter system model for the highways. Mass conservation model of a highway, characterized by $x \in [0, L]$, which is the position on the highway, is given by Equation 1,

$$\frac{\partial}{\partial t}\rho(x,t) = -\frac{\partial}{\partial x}q(x,t) \tag{1}$$

where $\rho(x,t)$ is the density of the traffic as a function of x, and time t, and $q(x,t)$ is the flow at given x, and t. The flow $q(x,t)$ is a function of $\rho(x,t)$, and the velocity v(x,t), as shown below in Equation 2.

$$q(x,t) = \rho(x,t)\,v(x,t) \qquad (2)$$

This model of a highway section is shown in Figure 4.1.

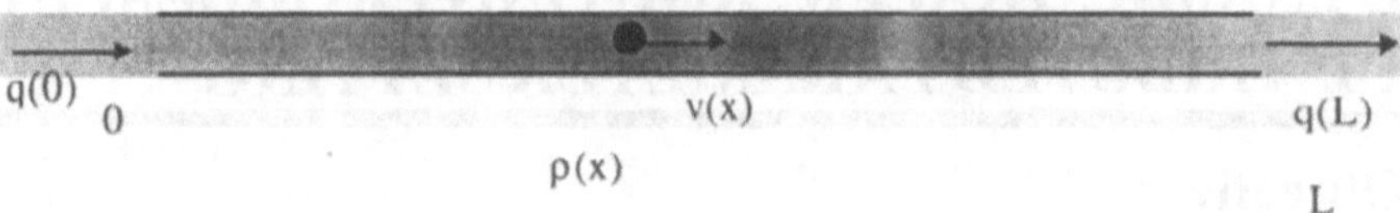


**Figure 4.1**: *Segment of Highway Model*

We use Greenshield model, which hypothesizes a linear relationship between v(x,t) and $\rho(x,t)$, as shown below

$$v = v_f\,(1 - \frac{\rho}{\rho_{max}}) \qquad (3)$$

where $v_f$ is the free flow speed, and $\rho_{max}$ is the jam density.

Let us consider the alternate route problem. Figure 4.2 below illustrates the problem.

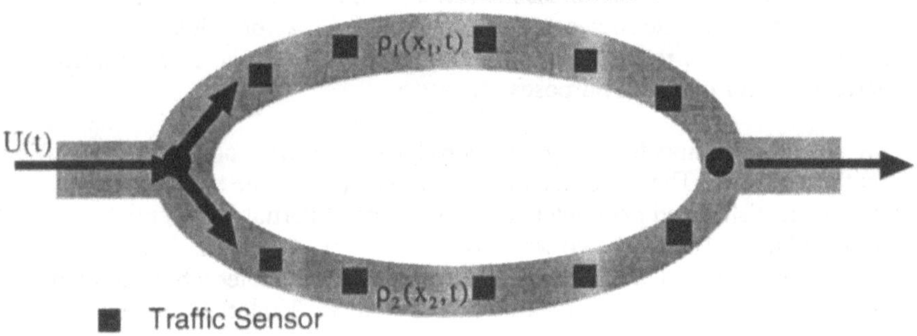


**Figure 4.2**: *Traffic Flow in Alternate Routes*

The following are the conservation equations for the two routes.

$$\frac{\partial}{\partial t}\rho_1(x_1,t) + \frac{\partial}{\partial x}q_1(x_1,t) = 0, \quad 0 \le x_1 \le L_1, \qquad (4)$$

$$\frac{\partial}{\partial t}\rho_2(x_2,t) + \frac{\partial}{\partial x}q_2(x_2,t) = 0, \quad 0 \le x_2 \le L_2, \qquad (5)$$

with the following boundary conditions at $(x_1, x_2) = (0,0)$.

$$q_1(0,t) = \beta(t)U(t) \qquad (6)$$

and $q_2(0,t) = (1 - \beta(t))U(t),\ 0 \le \beta(t) \le 1$

The boundary conditions show that the incoming traffic U(t) is divided between the two alternate routes, and that division is controlled by the variable β(t).

# 3. Sliding Mode Control

In this chapter, we will reduce the control problem in the PDE setting into one in Ordinary Differential Equation (ODE) setting. This will be shown in the next section. We will reduce the traffic routing problem to an equation of the form

$$\dot{x}(t) = f(x,t) + b(x,t)u(t) \tag{7}$$

where x(t) is a scalar function of time t, u(t) is the control variable same as β(t), and the functions f(x,t) and b(x,t) are some scalar functions. The state variable x(t) is equal to the difference in the travel times of the two routes. The detailed derivation of this will be given in the next section, where the meaning of all these terms will get clear. Our aim is to design the feedback control variable, i.e. come up with the function u=u(x) as a function of the state variable, so that x(t) goes to zero as t goes to infinity. In order to accomplish this, we will use a nonlinear control method called sliding mode control.

Sliding mode control is robust against uncertainties in the model. That means, if the actual system is different than the original system, the controller still provides good performance, provided that the disturbances meets certain criteria, as detailed below. To understand sliding mode control, let us consider the system given by:

$$\dot{x}(t) = -\operatorname{sgn}(x(t)) \tag{8}$$

where the signum function or the sign function sgn(.) is defined as

$$\operatorname{sgn}(x) = +1 \quad \text{if } x > 0 \tag{9}$$

$$\operatorname{sgn}(x) = -1 \quad \text{if } x < 0$$

If the initial value of x at time t=0 is positive, i.e. if x(0)>0, then the system becomes

$$\dot{x}(t) = -1 \tag{10}$$

which is same as

$$\frac{dx}{dt} = -1 \tag{11}$$

Integrating (11) gives us

$$\int_{x_0}^{X} dx = -\int_{0}^{T} dt \tag{12}$$

which gives us

$$x(t) = x_0 - t \tag{13}$$

This shows that starting at any positive initial value x, the trajectory moves towards x=0 at a rate –1unit/second. Hence, there is a finite time taken to reach the origin. We can plot the trajectory in two dimensional graph with x and t as the two axis (shown in Figure 4.3).

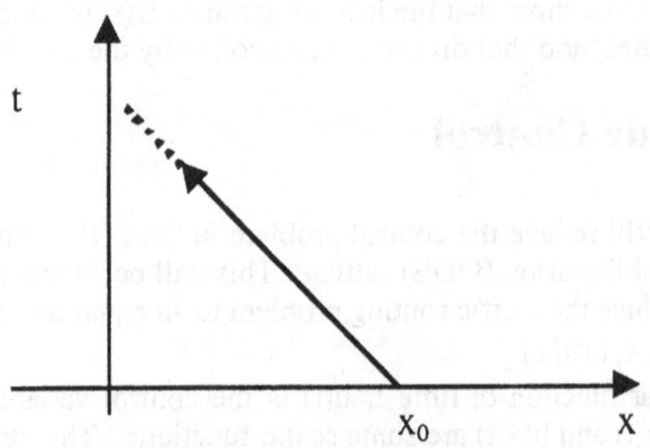


**Figure 4.3**: *Trajectory Starting at Positive x.*

If we show the same plot on the x-axis, we will have (Figure 4.4)

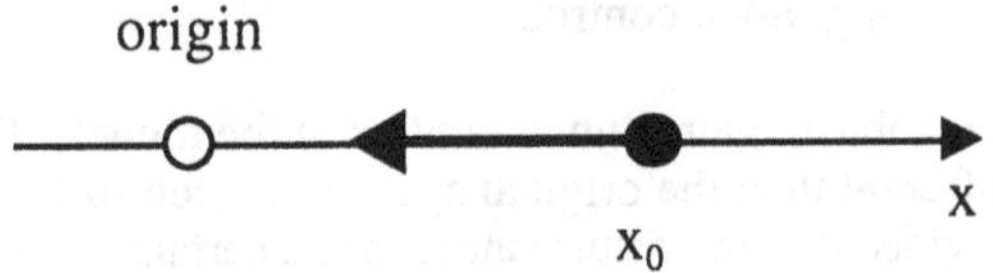


**Figure 4.4**: *Trajectory Starting at Positive x in one dimension.*

If the initial value of x at time t=0 is negative, i.e. if x(0)<0, then the system becomes

$$\dot{x}(t) = +1 \tag{14}$$

which is same as

$$\frac{dx}{dt} = +1 \tag{15}$$

Integrating (11) gives us

$$\int_{x_0}^{x} dx = \int_{0}^{T} dt \tag{16}$$

which gives us

$$x(t) = x_0 + t \tag{17}$$

This shows that starting at any negative initial value x, the trajectory moves towards x=0 at a rate 1unit/second. Hence, there is a finite time taken to reach the origin. We can plot the trajectory in two dimensional graph with x and t as the two axis (shown in Figure 4.5).

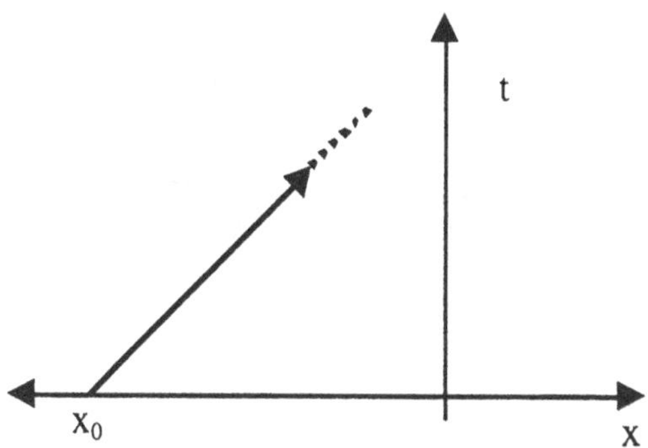


**Figure 4.5**: *Trajectory Starting at Negative x.*

If we show the same plot on the x-axis, we will have (Figure 4.6)

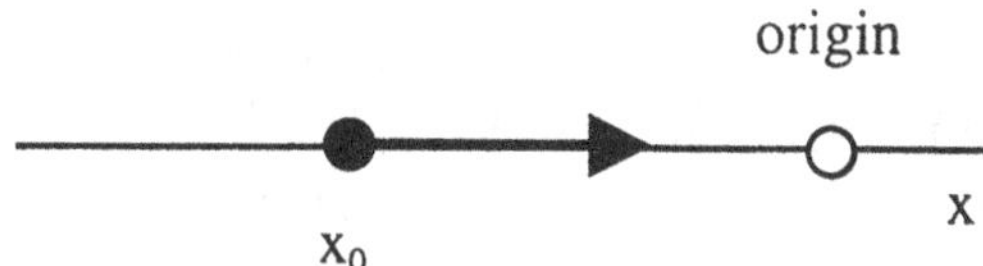


**Figure 4.6**: *Trajectory Starting at Negative x in one dimension.*

At the origin, the right hand side of the system (8) is discontinuous. Just on the right of the origin, the trajectory direction and magnitude (vector field) are given by what is shown in Figure 4 and just to the left, these are given by what is shown in Figure 4.6. This can lead to infinite switching at the origin.

In the classical sense, a solution to a differential equation of the form

$$\dot{x}(t) = f(x,t) \tag{18}$$

has a solution if f(t,x) is Lipschitz [1]. A function is called Lipschitz if it satisfies

$$\|f(x_1) - f(x_2)\| < k\|x_1 - x_2\| \tag{19}$$

The system (8) does not satisfy this condition since the right hand side is discontinuous. Therefore, there is a need for a new solution definition for systems like this which are discontinuous. This has been done by Fillipov [2, 3] extensively. He defined the solution in terms of *differential inclusions* where at the discontinuity we take a *convex hull* of the limiting vector fields. Convex hull is defined as the smallest convex set which contains the given members. For example convex hull of two given points is a straight line connecting the two points. Convex hull of three given points is a triangle connecting the three points.

Let us consider the following differential equation

$$\dot{\mathbf{x}} = f(\mathbf{x}, t) \tag{20}$$

where, f: $R^n X R \rightarrow R^n$ is essentially locally bounded and measurable. The solution of this differential equation is defined by Filippov by the following theorem.

Definition (Filippov's solution to differential equations with discontinuous right hand sides): A vector function **x**(.) is the solution of (18) on $[t_0,t_1]$ in the sense of Filippov, if **x**(.) is absolutely continuous on $[t_0,t_1]$, and for almost all t $\in$ $[t_0,t_1]$ it satisfies the following differential inclusion.

$$\dot{\mathbf{x}} \in K[f](\mathbf{x},t) \qquad (21)$$

There are two equivalent definitions for $K[f](\mathbf{x},t)$. The two definitions are described in [2, 4, 5, 6]. We will use one of the definitions here as.

$$K[f](x,t) \equiv \bigcap_{\delta>0} \bigcap_{\mu N=0} \overline{co}\, f(B(\mathbf{x},\delta) - N,t) \qquad (22)$$

where N denotes all sets of Lebesgue measure zero.

An alternate definition is based on a control representation of the system [7] as:

$$\dot{\mathbf{x}} = f(\mathbf{x},t,u_1(\mathbf{x},t),u_2(\mathbf{x},t),...u_p(\mathbf{x},t)) \qquad (23)$$

which at the discontinuities of $u_i(\mathbf{x},t)$ i=1,2,...p (all $u_i(\mathbf{x},t)$ being independent of each other), can be represented by the following differential inclusion.

$$\dot{\mathbf{x}} \in F(\mathbf{x},t,U_1(\mathbf{x},t),U_2(\mathbf{x},t),...U_p(\mathbf{x},t)) \qquad (24)$$

$U_i(\mathbf{x},t)$ i=1,2,....,p, being closed convex sets containing all the limit points of $u_i(\mathbf{x},t)$.

Essentially, Fillipov's solution states that the solution at the discontinuity is in some sense the average of the behavior on the different sides. For the example problem (8), this would mean that the trajectory would stay at the origin (equivalent to infinite switching about the origin). In practice however, there is digital implementation of the control which causes finite switching, since after any switching occurs, the trajectory stays on that side for some time and then switches (shown in Figure 4.7).

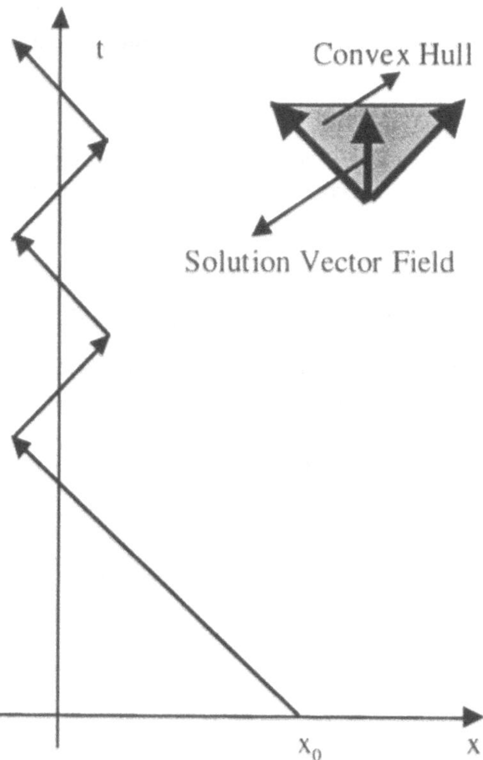


**Figure 4.7**: *Chattering*

In order to design a sliding mode control law to take the system from a starting point x(0) to a desired position value x=p, we define a sliding variable s(t) which is equal to the difference between the actual state and the desired state, i.e. s(t)=x(t)-p. In our case, the desired state can be x=0. For instance, when we want to drive the system error to zero, where the error could be the state variable, then the desired state is zero. We can force the system to go to zero value of s(t) if we have satisfy the following condition: when s(t)>0 then the derivative of s(t) with respect to time should be negative, and when s(t)<0 then the derivative of s(t) with respect to time should be positive. We can design a control law that accomplishes this. This is illustrated for a class of nonlinear systems next.

Let a single input nonlinear system be defined as

$$x^{(n)} = f(\mathbf{x},t) + b(\mathbf{x},t)u(t) \tag{25}$$

Here, $\mathbf{x}(t) = [x(t)\ \dot{x}(t)\ \ldots\ x^{(n-1)}]^T$ is the state vector, u is the control input and x is the output state. The superscript n on x(t) signifies the order of differentiation.

For example $x^{(1)}$ is used for $\dot{x}(t)$, $x^{(2)}$ is used for $\ddot{x}(t)$ and so on. In the equation, f(x,t) and b(x,t) are generally nonlinear functions of time and the states.

A time varying surface S(t) is defined by equating the variable s(t), defined below, to zero.

$$s(t) = \left(\frac{d}{dt} + \gamma\right)^{n-1} \tilde{x}(t) \tag{26}$$

Here, $\gamma$ is a constant, taken to be the bandwidth of the system, and $\tilde{x}(t) = x(t) - x_d(t)$ is the error in the output state where $x_d(t)$ is the desired output state.

Condition

$$\frac{1}{2}\frac{d}{dt}(s(t)^2) \leq -\eta|s(t)|,\ \eta > 0 \tag{27}$$

makes the surface S(t) an invariant set. All trajectories outside S(t) point towards the surface, and trajectories on the surface remain there. Equation (27) can also be written as

$$s(t)\dot{s}(t) \leq -\eta|s(t)|,\ \eta > 0 \tag{28}$$

which implies that for s(t)>0

$$\dot{s}(t) \leq -\eta \tag{29}$$

Integrating this equation from t=0 when $s=s_0$ to $t=t_r$, where $t_r$ is the time to reach the surface S(t), yields

$$\int_{s_0}^{0} ds \leq -\int_{0}^{t_r} \eta dt \tag{30}$$

which implies

$$t_r \leq (s_0/\eta) \tag{31}$$

A similar result is obtained when $s_0<0$, so that

$$t_r \leq (|s_0|/\eta) \tag{32}$$

This shows that it takes finite time to reach the surface S(t) from outside. Moreover the definition (26) implies that once the surface is reached, the convergence to zero error is exponential as shown in Figure 4.8.

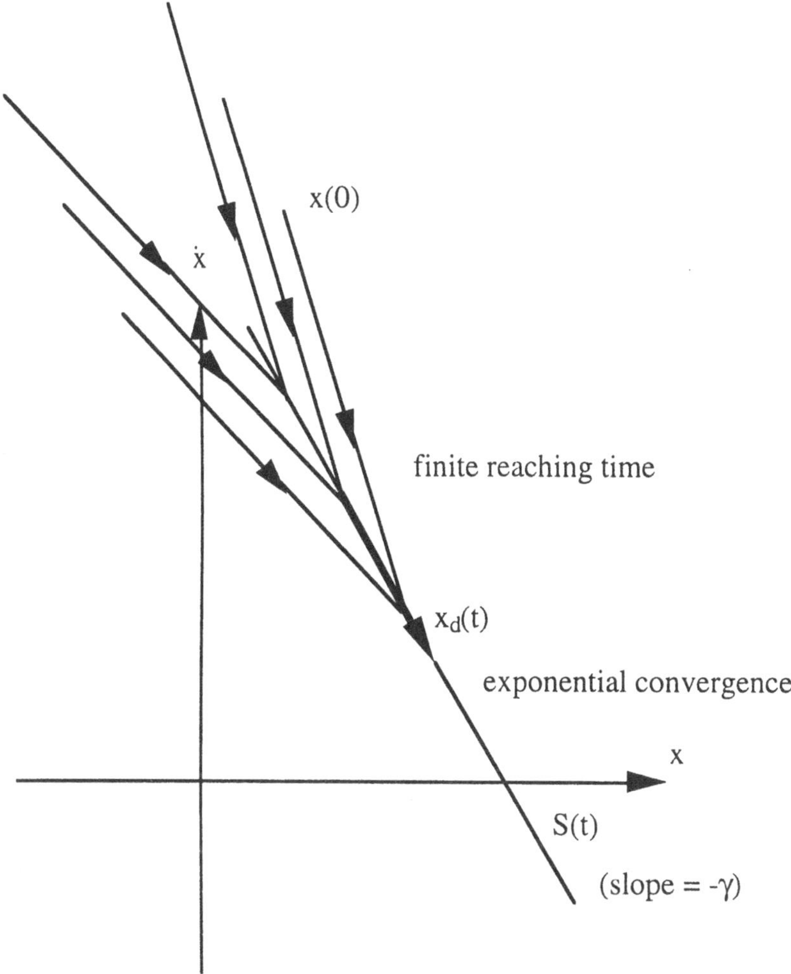


**Figure 4.8**: *Sliding Mode in Phase Plane*

Chattering is caused by non-ideal switching around the switching surface. Delay in digital implementation causes s(t) to pass to the other side of the surface, which in turn produces chattering. Chattering can be best explained by a hysteristic behavior near the switching surface for a first order system as shown in Figure 4.9. The two values of $\dot{s}(t)$ are $g^+$ and $g^-$, $g^+$ being the value of $\dot{s}(t)$ near S(t) when s(t) is positive, and $g^-$ being the value of $\dot{s}(t)$ near S(t) when s(t) is negative. It can be seen that decreasing, which in turn means increasing the sampling rate, increases the frequency of chattering. Chattering in the s(t) variable is shown in Figure 4.10. Chattering is undesirable, for it can excite the higher order unmodeled dynamics of the system and produce undesired performance.

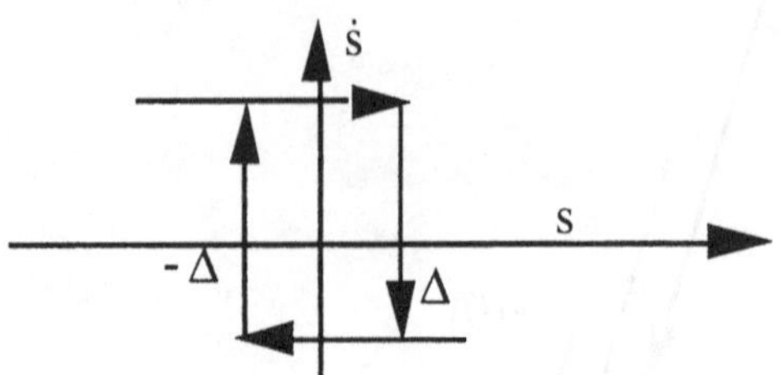


**Figure 4.9**: *Hysterisis in sliding mode*

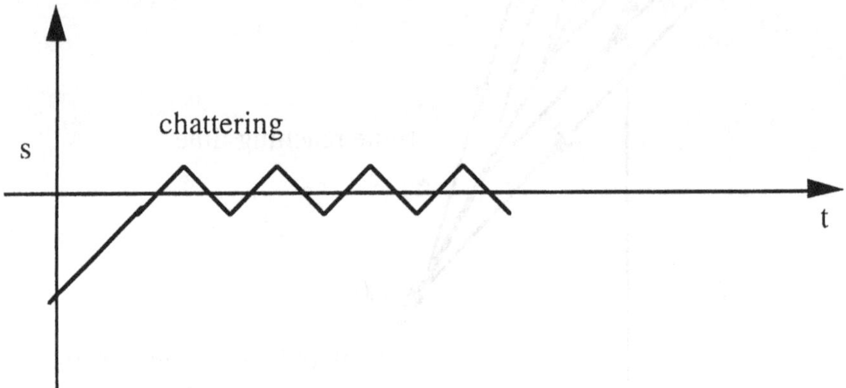


**Figure 4.10**: *Chattering in the sliding variable*

Consider a second order system [8, 9, 10, 11]

$$\ddot{x}(t) = f(\mathbf{x},t) + u(t) \tag{33}$$

where f(**x**,t) is generally nonlinear and/or time varying and is estimated as $\hat{f}(\mathbf{x},t)$, u(t) is the control input, and x(t) is the state to be controlled so that it follows a desired trajectory $x_d(t)$. The estimation error on f(x,t) is assumed to be bounded by some known function F=F(**x**,t), so that

$$|\hat{f}(\mathbf{x},t) - f(\mathbf{x},t)| \leq F(\mathbf{x},t). \tag{34}$$

We define a sliding variable according to (26)

$$s(t) = (\frac{d}{dt} + \gamma)\tilde{x}(t) = \dot{\tilde{x}}(t) + \gamma\tilde{x}(t) \tag{35}$$

Differentiation of the sliding variable yields

$$\dot{s}(t) = \ddot{x}(t) - \ddot{x}_d(t) + \gamma\dot{\tilde{x}}(t) \tag{36}$$

Substituting Equation (33) in Equation (36), we have

$$\dot{s}(t) = f(\mathbf{x},t) + u(t) - \ddot{x}_d(t) + \gamma\dot{\tilde{x}}(t) \tag{37}$$

The approximation of control law $\hat{u}(t)$ to achieve $\dot{s}(t) = 0$ is

$$\hat{u}(t) = -\hat{f}(\mathbf{x},t) + \ddot{x}_d(t) - \gamma\dot{\tilde{x}}(t) \tag{38}$$

To account for the uncertainty in f while satisfying the sliding condition (27), take the control law as:

$$u(t) = \hat{u}(t) - k(\mathbf{x},t)\ \mathrm{sgn}(s(t)) \tag{39}$$

Choosing

$$k(\mathbf{x},t) = F(\mathbf{x},t) + \eta \tag{40}$$

ensures the invariant condition of Equation (27), since

$$\frac{1}{2}\frac{d}{dt}(s(t)^2) = \dot{s}(t)s(t) = [f(\mathbf{x},t) - \hat{f}(\mathbf{x},t) - k(\mathbf{x},t)\ \mathrm{sgn}(s(t))]s(t \tag{41}$$

$$= (f(\mathbf{x},t) - \hat{f}(\mathbf{x},t))s(t) - k(\mathbf{x},t)|s(t)|$$

Equations (37) and (39) were used to derive the above result. Hence, by using (39), we ensure that for the system (33), the system trajectory will take finite time to reach the surface S(t), after which the errors will exponentially go to zero.

Now, consider the system

$$\ddot{x}(t) = f(\mathbf{x},t) + b(\mathbf{x},t)u(t) \tag{42}$$

where $b(\mathbf{x},t)$ is bounded as

$$0 \le b_{min}(\mathbf{x},t) \le b(\mathbf{x},t) \le b_{max}(\mathbf{x},t) \tag{43}$$

The control gain $b(\mathbf{x},t)$ and its bounds can be time varying or state dependent. Since the control input is multiplied by the control gain in the dynamics, the geometric mean of the lower and upper bounds of the gain is a reasonable estimate:

$$\hat{b}(\mathbf{x},t) = \sqrt{b_{min}(\mathbf{x},t)b_{max}(\mathbf{x},t)} \tag{44}$$

The bounds (43) can also be written as

$$\alpha^{-1}(\mathbf{x},t) \le \hat{b}(\mathbf{x},t)/b(\mathbf{x},t) \le \alpha(\mathbf{x},t) \tag{45}$$

where

$$\alpha(\mathbf{x},t) = \sqrt{b_{max}(\mathbf{x},t)/b_{min}(\mathbf{x},t)} \tag{46}$$

Take the control law as

$$u(t) = \hat{b}(\mathbf{x},t)^{-1}[\hat{u}(t) - k(\mathbf{x},t)\ \mathrm{sgn}(s(t))] \tag{47}$$

where

$$k(\mathbf{x},t) \ge \alpha(\mathbf{x},t)(F(\mathbf{x},t) + \eta) + (\alpha(\mathbf{x},t) - 1)|\hat{u}(t)| \tag{48}$$

It is shown below that this control law ensures the sliding condition (27). First, we note that

$$\dot{s}(t) = f(\mathbf{x},t) + b(\mathbf{x},t)u(t) - \ddot{x}_d(t) + \gamma\dot{\tilde{x}}(t) \tag{49}$$

Using the control u(t) from Equation (47) we obtain

$$s(t)s(t) = [(f(\mathbf{x},t) - b(\mathbf{x},t)\hat{b}(\mathbf{x},t)^{-1}\hat{f}) + (1 - b(\mathbf{x},t)\hat{b}(\mathbf{x},t)^{-1})(-\ddot{x}_d(t) + \gamma\dot{\tilde{x}}(t)) \tag{50}$$

$$- b(\mathbf{x},t)\hat{b}(\mathbf{x},t)^{-1}k(\mathbf{x},t)\ \mathrm{sgn}(s(t))]s(t)$$

From Equation (47), in order to satisfy (27) the sliding gain k must be such that

$$k(\mathbf{x},t) \ge |\ b(\mathbf{x},t)^{-1}\hat{b}(\mathbf{x},t)f(\mathbf{x},t) - \hat{f}(\mathbf{x},t)) \tag{51}$$

$$+ (b(\mathbf{x},t)^{-1}\hat{b}(\mathbf{x},t)- 1)(-\ddot{x}_d(t) + \gamma\dot{\tilde{x}}(t))| + \eta b(\mathbf{x},t)^{-1}\hat{b}(\mathbf{x},t)$$

Using Equations (34) and (45) in Equation (29), we obtain

$$k(\mathbf{x},t) \ge |\hat{b}(\mathbf{x},t)b(\mathbf{x},t)^{-1} - 1||\ \hat{f}(\mathbf{x},t) - \ddot{x}_d(t) + \gamma\dot{\tilde{x}}(t)\ | + \eta\hat{b}(\mathbf{x}, \tag{52}$$

$$+ F(\mathbf{x},t)\hat{b}(\mathbf{x},t)b(\mathbf{x},t)^{-1}$$

which leads to (48).

# 4. Chattering Reduction

There are essentially two ways of producing chattering free performance in sliding mode. One technique for producing chattering free sliding mode [12, 13, 14] utilizes higher order sliding mode. In that technique the state equation is differentiated to produce a differential equation, which consists of the derivative of the control input, which then is utilized as a new control variable. Hence, this new control variable can be discontinuous while still producing a continuous control input. The difficulty with this technique is that the derivative of the state variable (which is differentiated to produce the derivative of the control input in the dynamic equation) is not available for measurement, and hence observers have to be designed to estimate that variable.

The other approach for chattering reduction is based on introducing a boundary layer around the switching surface and using a continuous control within the boundary layer, keeping the boundary layer attractive to the trajectories outside the boundary layer [9, 10, 11]. In sliding mode control, the term in the control law to counter the system uncertainties is given by $k(\mathbf{x},t)\mathrm{sgn}(s(t))$, where $k(\mathbf{x},t)$ is the discontinuity gain expressed as a function of the state $\mathbf{x}(t)$ and time t, s(t) is the sliding variable. In the method proposed in [9, 10, 11] , this term is replaced by $k(\mathbf{x},t)\mathrm{sat}(s(t),\phi)$, where $\phi$ is the boundary layer thickness which is made varying in order to take advantage of the system bandwidth. The function sat $(s(t),\phi)$ is defined as

$$\begin{aligned} \mathrm{sat}(s(t),\phi) &= s(t)/\phi && \text{if } |s(t)| < \phi \\ \mathrm{sat}(s(t),\phi) &= \mathrm{sgn}(s(t)) && \text{otherwise} \end{aligned} \tag{53}$$

In order to utilize the bandwidth of the system, it is not necessary to vary $\phi$. One drawback of varying the boundary layer is that for some systems the boundary width can become large. An alternate method for chattering reduction is proposed [15, 16, 17, 18] that achieves the same results without varying the boundary layer so that the disadvantage of the varying boundary is overcome. For some systems, as shown in this chapter, the sat function does not give satisfactory results and hence some other functions should be used which provide favorable results. New functions are proposed inside the boundary layer, which not only reduce chattering, but also cause error convergence for the system. The function to be used inside the boundary layer is determined by the dynamics of the system.

To remove chattering, a thin boundary of thickness $\phi$ around the switching surfaceis defined as

$$B(t) = \{\mathbf{x}(t), |s(t)| < \phi\} \tag{54}$$

We can guarantee that all the trajectories outside the boundary layer are attracted towards the boundary if the distance to the boundary layer always decreases. This is true when

$$s(t) \geq \phi(t) \Rightarrow \frac{d}{dt}(s(t) - \phi(t)) \leq -\eta \text{ i.e } \dot{s}(t) \leq -\eta + \dot{\phi}(t) \tag{55}$$

$$s(t) \leq -\phi(t) \Rightarrow \frac{d}{dt}(s(t) + \phi(t)) \geq \eta \text{ i.e } \dot{s}(t) \geq \eta - \dot{\phi}(t)$$

Combining the above equations, we can write this condition as

$$|s(t)| \geq \phi(t) \Rightarrow \frac{1}{2}\frac{d}{dt}s(t)^2 \leq (\dot{\phi}(t) - \eta)|s(t)| \tag{56}$$

Notice that the boundary layer attraction condition is more stringent during boundary layer contraction ($\dot{\phi} < 0$) and less stringent during boundary layer expansion ($\dot{\phi} > 0$). To satisfy (56), for the system (33), we define

$$\bar{k}(\mathbf{x},t) = k(\mathbf{x},t) - \dot{\phi}(t) \tag{57}$$

and use the control law

$$u(t) = \hat{u}(t) - \bar{k}(\mathbf{x},t)\text{sat}(s(t),\phi(t)) \tag{58}$$

Inside the boundary, the system trajectories can be expressed in terms of the variable s as

$$\dot{s}(t) = -\bar{k}(\mathbf{x},t)s(t)/\phi(t) - \Delta f(\mathbf{x},t) \tag{59}$$

where $\Delta f(\mathbf{x},t) = \hat{f}(\mathbf{x},t) - f(\mathbf{x},t)$. Since $\Delta f(\mathbf{x},t)$ and $\bar{k}(\mathbf{x},t)$ are continuous, we can express (59) as

$$\dot{s}(t) = -\bar{k}(\mathbf{x}_d,t)s(t)/\phi(t) + (-\Delta f(\mathbf{x}_d,t) + o(\xi)) \tag{60}$$

where, $o(\xi)$ represents the error terms introduced by replacing $\mathbf{x}(t)$ by $\mathbf{x}_d(t)$ in the first two terms in Equation (59). The variable s(t) can be viewed as the output of a first order low pass filter with bandwidth __ if we let

$$\bar{k}(\mathbf{x}_d,t)/\phi(t) = \gamma \tag{61}$$

This filter removes the high frequency chattering to give a smooth s(t). The bandwidth should be small as compared to high frequency unmodeled dynamics. Using the definition (57) in Equation (61), we obtain the variation of $\phi(t)$ with time in terms of the following differential equation:

$$\dot{\phi}(t) = -\gamma\phi(t) + k(\mathbf{x}_d,t). \tag{62}$$

The expression for sliding gain is obtained by using (57) in the above equation.

$$\bar{k}(\mathbf{x},t) = \gamma\phi(t) + k(\mathbf{x},t) - k(\mathbf{x}_d,t) \tag{63}$$

Figure 4.11 shows the first order low pass filter for s(t), where p is the Laplace operator[2] d/dt.

---

[2]Standard notation for this would be a complex variable s used in Laplace transforms, but symbol s has already been used as the sliding variable.

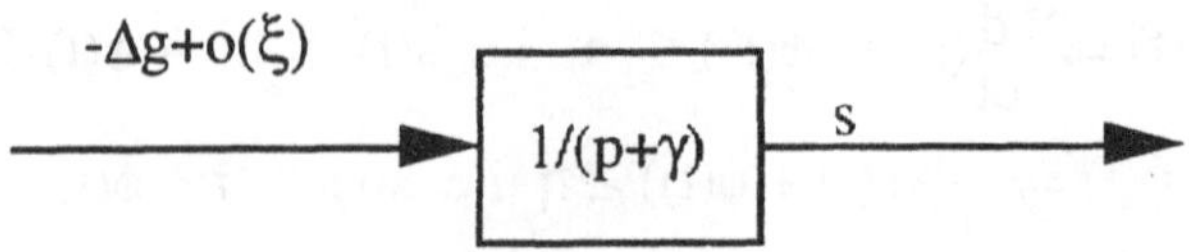


**Figure 4.11**: *Filter for chattering reduction*

The same filter can be obtained by using a constant width boundary such that

$$\phi(t) = \text{constant} \tag{64}$$

$$\dot{\phi}(t) = 0$$

By taking a constant boundary width, the condition (56) is reduced to condition (27). From (57), we see that

$$\bar{k}(\mathbf{x},t) = k(\mathbf{x},t) \tag{65}$$

Now, take the control law as

$$u(t) = \hat{u}(t) - k(\mathbf{x},t)\ \mathrm{msat}(a(\mathbf{x},t),s(t),\phi) \tag{66}$$

where the modified saturation function msat($a(\mathbf{x},t),s(t),\phi$) is defined as

$$\mathrm{msat}(a(\mathbf{x},t),s(t),\phi) = a(\mathbf{x},t)s(t)/\phi \quad \text{for } |s(t)| < \phi \tag{67}$$

$$\mathrm{msat}(a(\mathbf{x},t),s(t),\phi) = \mathrm{sgn}(s(t)) \quad \text{otherwise} \tag{68}$$

Using the control law (43), the dynamics inside the boundary layer becomes

$$\dot{s}(t) = -k(\mathbf{x}_d,t)as(t)/\phi + (-\Delta f(\mathbf{x}_d,t) + o(\xi)) \tag{69}$$

We can assign appropriate values for the variable $a(\mathbf{x},t)$ in order to achieve a desirable bandwidth

$$a(\mathbf{x},t) = \frac{\gamma\phi}{k(\mathbf{x}_d,t)} \tag{70}$$

The above value gives the same filter as can be obtained by using the sat function. The advantage of using the msat function is that the boundary width is kept fixed so that the area in which the system trajectories are attracted towards the boundary is not changed. On the other hand, the boundary width can become large by using the sat function as is shown in the next section. The msat function produces the same filter as the sat function by changing the variation of width of the boundary layer into a variation of height as shown in Figure 4.12. Notice that the msat function is discontinuous at $s(t)= \phi$ If the trajectories on both sides of the boundary face inwards i.e. towards S(t), the discontinuity does not produce any problems. This is the case when the input $-\Delta f(\mathbf{x}_d) + o(\xi)$ to the first order filter is an impulse input.

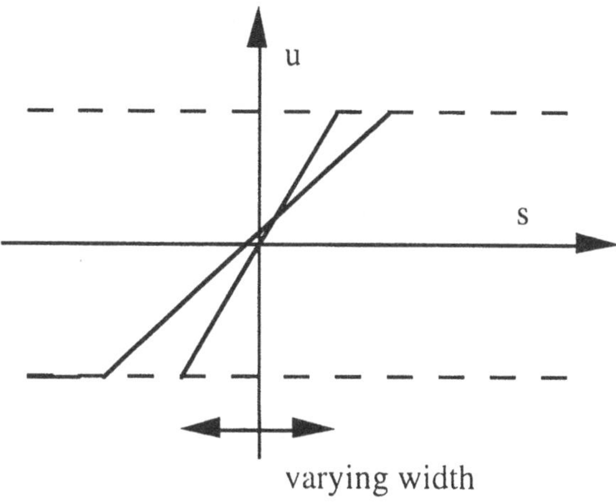


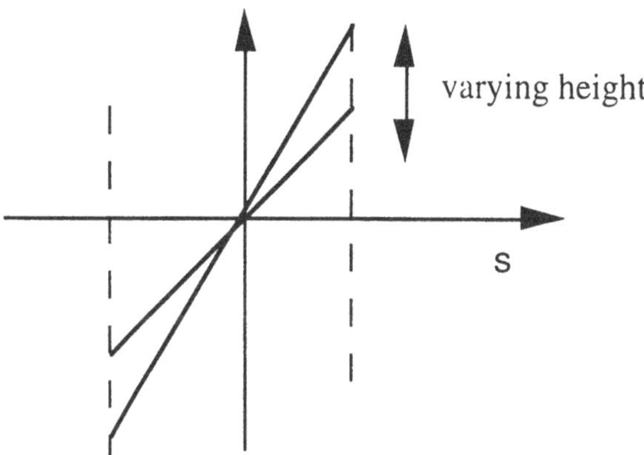


**Figure 4.12** : *Interpolation in the boundary layer*

Now if the input $-\Delta f(\mathbf{x}_d) + o(\xi)$ to the first order filter is a step input, then the variable s(t) has a steady state value. Similarly, if the input term is a ramp, then s(t) keeps increasing. In that case, if the sat function with a varying boundary is being used, the boundary might keep increasing too. When a fixed boundary width is used, the variation of s with respect to time may increase until it hits the boundary layer and once it is out of the boundary, it is forced back inwards because of the attractiveness of the boundary layer. This effect causes chatteringon the boundary as shown in Figure 4.13. This chattering is caused due to the discontinuity in the msat function at $s(t)= \phi$ . The amount of discontinuity is governed by the variable a, which in turn fixes the bandwidth of the s-filter inside the boundary. Therefore, the amount of discontinuity limits the achievable bandwidth of the filter. This problem is solved by forcing the trajectories on both sides of the boundary to face inwards. To accomplish that, an integral action is needed, as explained next.

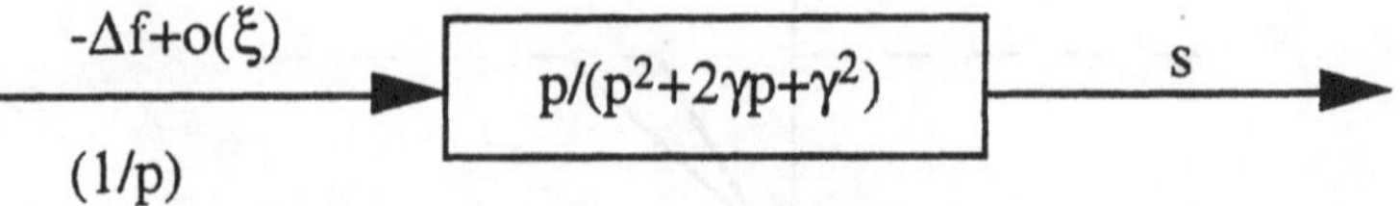


**Figure 4.13**: *Integral filter for chattering reduction*

To properly handle the step input of the filter of Figure 11, we need to introduce an integral action within the boundary layer. When $-\Delta f(\mathbf{x}_d,t) + o(\xi)$ is constant, so that its Laplace transform can be expressed as m/p, where m is a constant, we modify the control law to

$$u(t) = \hat{u}(t) - k(\mathbf{x},t)\ \mathrm{int}(a(\mathbf{x},t),j(\mathbf{x},t),s(t),\phi) \tag{71}$$

where

$$\mathrm{int}(a(\mathbf{x},t),j(\mathbf{x},t),s(t),\phi) = a(\mathbf{x},t)s(t)/\phi + \frac{j(\mathbf{x},t)}{\phi}\int_0^t s(t)dt \quad \text{for } |s(t)| \leq \phi$$

$$\mathrm{int}\ (a(\mathbf{x},t),j(\mathbf{x},t),s(t),\phi) = \mathrm{sgn}(s(t)) \qquad \text{otherwise.} \tag{72}$$

Using control law (71), the dynamics within the boundary layer can be approximated as

$$\dot{s}(t) = -\bar{k}(\mathbf{x}_d,t)a(\mathbf{x},t)s(t)/\phi - \bar{k}(\mathbf{x}_d,t)\frac{j(\mathbf{x},t)}{\phi}\int_0^t s(t)dt + (-\Delta f(\mathbf{x}_d,t) + o(\xi)) \tag{73}$$

To assign the values for a and j, we can either have a fixed boundary and take $a(\mathbf{x},t) = 2\gamma\phi/k(\mathbf{x}_d,t)$ and $j(\mathbf{x},t) = \gamma^2\phi/k(\mathbf{x}_d,t)$, or we can have a varying boundary by taking $a(x,t) = 1$, $\bar{k}(\mathbf{x}_d,t)/\phi(t) = 2\gamma$ and $j(\mathbf{x},t) = \gamma^2\phi(t)/k(\mathbf{x}_d,t)$. In the latter case, the boundary width varies according to

$$\dot{\phi}(t) + 2\gamma\phi(t) = k(\mathbf{x}_d,t) \tag{74}$$

The filter obtained for the constant and the time varying $\phi$ is the same. The filter is shown in Figure 4.13. The presence of p in the numerator of the filter transfer function lets s(t) converge to zero when the input is constant. By using the integral action, there is no chattering across the boundary, and the trajectories on both sides of the boundary are directed towards S(t).

Extending the argument in the same vein, if the filter input has a term with Laplace transform $m/p^n$, we need to introduce n integral terms. For instance, for a system with

$$\Delta f(\mathbf{x}) = -2.5x^2(t) \text{ and}$$
$$x_d(t) = 0.23t \tag{75}$$

the control law inside the boundary layer should have integrals up to the third order.

Now, for the system (42), in order to satisfy (55) in the presence of uncertainty _ on the control gain, for a variable width saturation function implementation, we let

$$\dot{\phi}(t) > 0 \Rightarrow \bar{k}(\mathbf{x},t) = k(\mathbf{x},t) - \dot{\phi}(t)/\alpha(\mathbf{x},t) \qquad (76)$$

$$\dot{\phi}(t) < 0 \Rightarrow \bar{k}(\mathbf{x},t) = k(\mathbf{x},t) - \dot{\phi}(t)\alpha(\mathbf{x},t)$$

The balance condition (61) for this system can be written as

$$\left[\frac{\bar{k}(\mathbf{x}_d,t)}{\phi(t)}\right]\left[\frac{b(\mathbf{x}_d,t)}{\hat{b}(\mathbf{x}_d,t)}\right]_{max} = \gamma \qquad (77)$$

or

$$\bar{k}(\mathbf{x}_d,t) = \gamma\phi(t)/\alpha(\mathbf{x}_d,t) \qquad (78)$$

Applying relation (78) to (76) yields

$$\dot{\phi}(t) > 0 \Rightarrow \gamma\phi(t)/\alpha(\mathbf{x}_d,t) = k(\mathbf{x}_d,t) - \dot{\phi}(t)/\alpha(\mathbf{x}_d,t) \qquad (79)$$

$$\dot{\phi}(t) < 0 \Rightarrow \gamma\phi(t)/\alpha(\mathbf{x}_d,t) = k(\mathbf{x}_d,t) - \dot{\phi}(t)\alpha(\mathbf{x}_d,t)$$

that is,

$$k(\mathbf{x}_d,t) \geq \gamma\phi(t)/\alpha(\mathbf{x}_d,t) \Rightarrow \dot{\phi}(t) + \gamma\phi(t) = \alpha(\mathbf{x}_d,t)k(\mathbf{x}_d,t) \qquad (80)$$

$$k(\mathbf{x}_d,t) < \gamma\phi(t)/\alpha(\mathbf{x}_d,t) \Rightarrow \dot{\phi}(t) + \gamma\phi(t)/[\alpha(\mathbf{x}_d,t)]^2 = k(\mathbf{x}_d,t)/\alpha(\mathbf{x}_d,t)$$

with initial condition

$$\phi(0) = \alpha(\mathbf{x}_d,0)k(\mathbf{x}_d(0),0)/\gamma \qquad (81)$$

The sliding gain for the sat function can be obtained as

$$\bar{k}(\mathbf{x},t) = (\bar{k}(\mathbf{x},t) - \bar{k}(\mathbf{x}_d,t)) + \bar{k}(\mathbf{x}_d,t) = (k(\mathbf{x},t) - k(\mathbf{x}_d,t)) + \phi(t)\gamma/\alpha(\mathbf{x}_d,t) \qquad (82)$$

We can eliminate the variation of $\phi(t)$ and use the msat function. Since, for this case, $\dot{\phi}(t) = 0$, we obtain $\bar{k}(\mathbf{x},t) = k(\mathbf{x},t)$. The balance condition is changed to

$$\left[\frac{k(\mathbf{x}_d,t)a(\mathbf{x},t)}{\phi}\right]\left[\frac{b(\mathbf{x}_d,t)}{\hat{b}(\mathbf{x}_d,t)}\right]_{max} = \gamma \qquad (83)$$

To obtain a low pass filter of bandwidth $\gamma$, the variable a will be taken as

$$a(\mathbf{x},t) = \frac{\gamma\phi}{\alpha(\mathbf{x}_d,t)k(\mathbf{x}_d,t)} \qquad (84)$$

Implementation of this scheme is simpler because $\phi$ is constant and the design of a(x,t) is straightforward.

For system (42), the system trajectories inside the boundary layer can be expressed as

$$s(t) = (f(\mathbf{x},t) - b(\mathbf{x},t)\hat{b}(\mathbf{x},t)^{-1}\hat{f}(\mathbf{x},t)) + (1 - b(\mathbf{x},t)\hat{b}(\mathbf{x},t)^{-1})(-x_d(t) + \gamma\dot{\tilde{x}}(t)) - b(\mathbf{x},t)\hat{b}(\mathbf{x},t)^{-1}k(\mathbf{x},t)\ \mathrm{sgn}(s(t)) \qquad (85)$$

which can be rewritten as

$$\dot{s}(t) = - b(\mathbf{x},t)\hat{b}(\mathbf{x},t)^{-1}k(\mathbf{x},t)\ \mathrm{sgn}(s(t)) - i(\mathbf{x},t) \qquad (86)$$

where

$$i(\mathbf{x},t) = -(f(\mathbf{x},t) - b(\mathbf{x},t)\hat{b}(\mathbf{x},t)^{-1}\hat{f}(\mathbf{x},t)) - (1 - b(\mathbf{x},t)\hat{b}(\mathbf{x},t)^{-1})(- \qquad (87)$$

The input to the filter for s(t) variable is $-i(\mathbf{x}_d,t)$ and therefore this term should be analyzed as explained for system (33) to ascertain which function should replace the sgn function for chattering reduction and error convergence. For the system with uncertain input functions, there is an additional term of $b(\mathbf{x},t)(\hat{g}(\mathbf{x},u,t)-g(\mathbf{x},u,t))$ in the $i(\mathbf{x},t)$ term.

## 5. Numerical Examples

Consider a second order system of the form (33)

$$\ddot{x} = -|\sin t|\dot{x}^2\cos 3x + u \tag{88}$$

The control law (39) for this example is

$$u = 1.5\dot{x}^2\cos 3x - (\pi^2/4)\sin(\pi t/2) - 20\dot{\tilde{x}} - (0.1 + 0.5\dot{x}^2|\text{co} \tag{89}$$

where the s variable is defined in (35). Figure 4.14 shows that tracking performance using this control law is excellent but at a price of high control chattering.

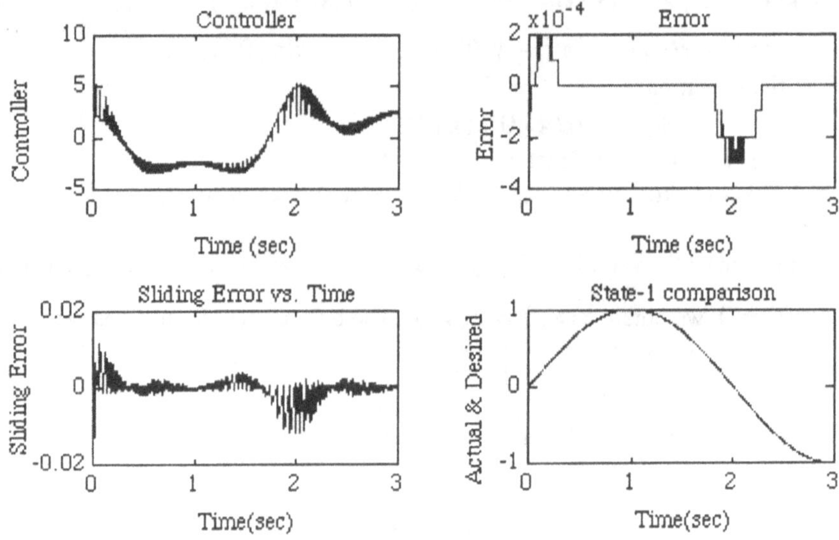


**Figure 4.14:** *Simulation results using signum function*

To remove chattering, we define a constant width boundary $\phi = 0.1$ and use the control law

$$u(t) = \hat{u}(t) - k(\mathbf{x},t)\text{sat}(s(t)/\phi)$$

or

$$u = 1.5\dot{x}^2\cos 3x - (\pi^2/4)\sin(\pi t/2) - 20\dot{\tilde{x}} - (0.1 + \tag{90}$$

As is evident from Figure 4.15, although the error has increased, the performance is still acceptable and chattering has been removed to smoothen the output.

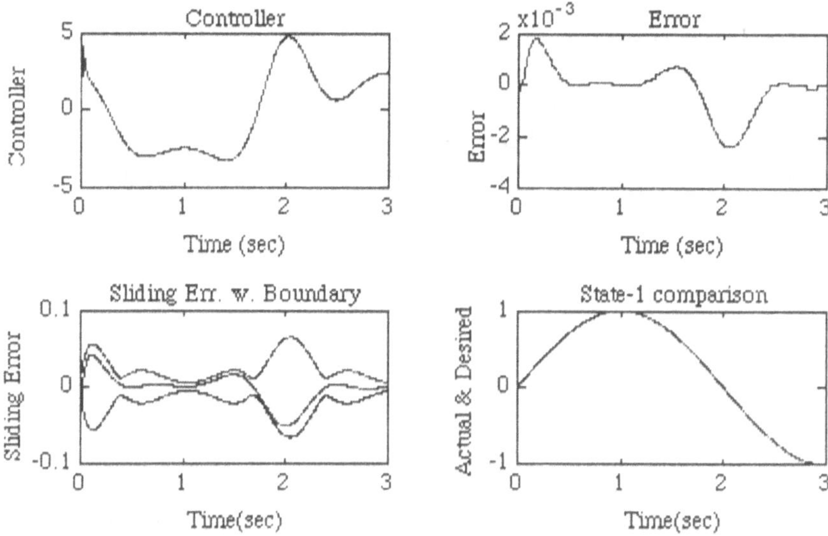


**Figure 4.15**: *Simulation results using sat function*

Now, the boundary $\phi$ is made varying and the control law (58) is used, which for this example is

$$u = 1.5\dot{x}^2\cos 3x - (\pi^2/4)\sin(\pi t/2) - 20\bar{x} - (0.1 + 0.5\dot{x}^2|\cos 3x| - \phi)\mathrm{sat}(s/\phi) \tag{91}$$

with

$$\dot{\phi} = -20\phi + 0.5\dot{x}_d^2|\cos 3x_d| + 0.1$$

Figure 4.16 shows that by using a varying boundary, the error has improved and at the same time there is no chattering.

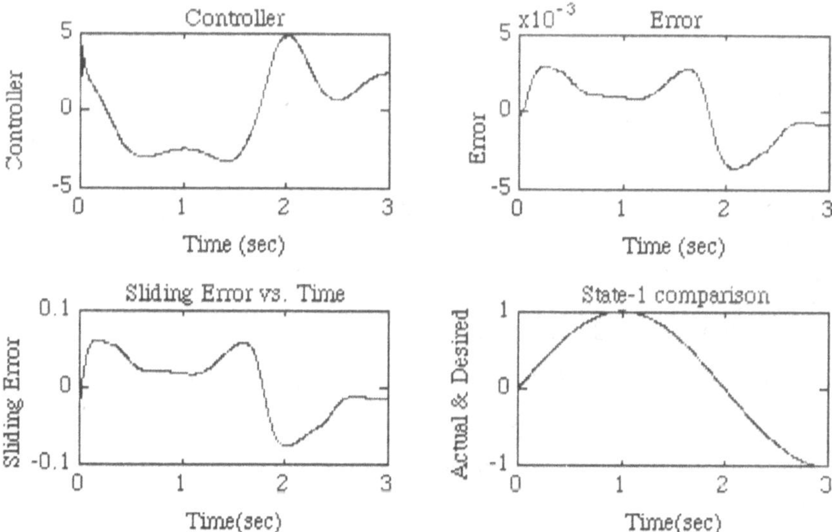


**Figure 4.16**: *Simulation results using variable width sat function*

Now the boundary is kept constant at 0.1, which is greater than the maximum value of the varying boundary for the previous case. The control law (66) is used as

$$u = 1.5\dot{x}^2\cos 3x - (\pi^2/4)\sin(\pi t/2) - 20\dot{\tilde{x}} - (0.1 + 0.5\dot{x}^2|\cos 3x|)\text{msat}(a,s,0.1) \tag{92}$$

with

$$a = 20(0.1)/(0.5\dot{x}_d^2|\cos 3x_d|) \tag{93}$$

We obtain the same results as were obtained by using the control (69) and (70), as is shown in Figure 4.17.

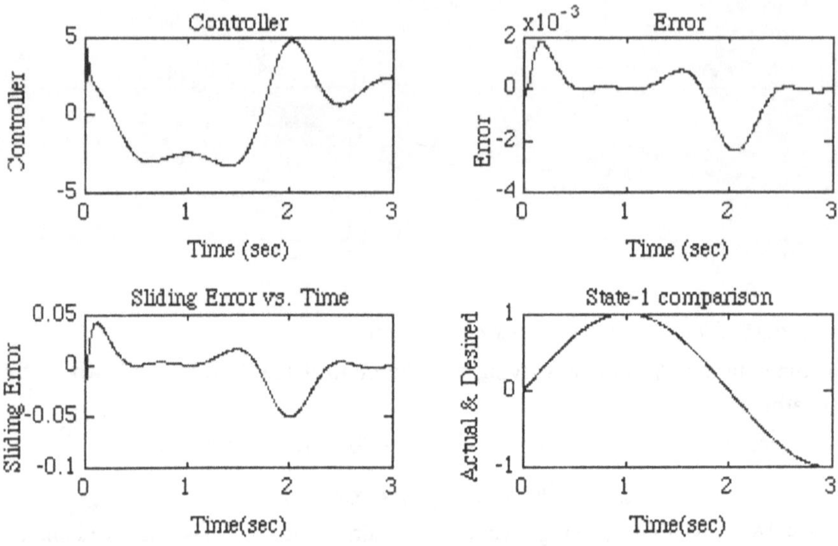


**Figure 4.17:** *Simulation results using msat function*

Now, to show the advantage of using the new interpolated functions with fixed boundaries, consider a second order system of the form (33) with simulation parameters and the desired output as

$$f = -2.00;\ \hat{f} = -1.0; F = 1.01;\ \eta = 0.1;\ \gamma = 20;\ x_d = \sin(\pi t/2) \tag{94}$$

Define a term form-n to indicate that the Laplace transform of the input to the filter inside the boundary layer is $m/p^n$. For example, for a step input, the system will be of form-1; and for a ramp input it will be of form-2 and so on. To reduce chattering, a time varying boundary is used and the control law (58) with parameters (94) is applied. The control input is

$$u = 1 - (\pi^2/4)\sin(\pi t/2) - 20\dot{\tilde{x}} - (1.11 - \dot{\phi})\text{sat}(s/\phi) \tag{95}$$

$$\dot{\phi} = -20\phi + 1.11 \tag{96}$$

and a steady state error is obtained in the value of s and the output error, as shown in Figure 4.18.

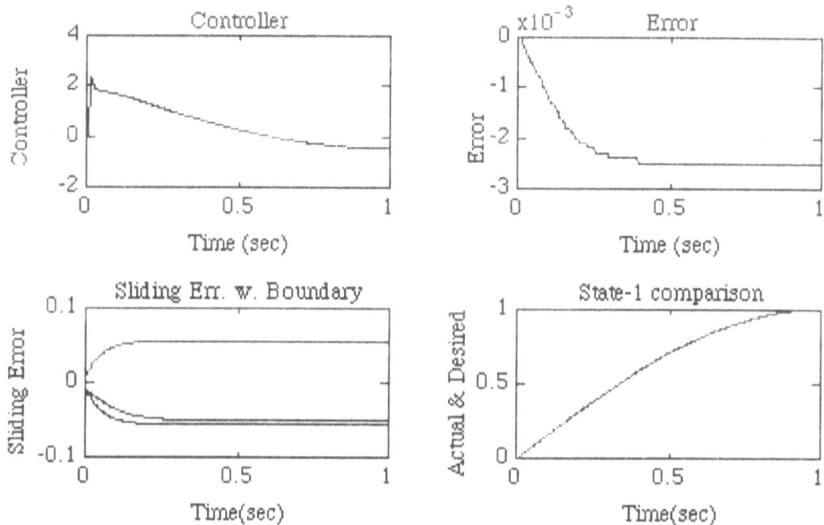


**Figure 4.18** *Simulation results on form-1 system using variable width sat function*

Consider a system of form-2 with the following simulation parameters and desired output.

$$f = -2.00t;\ \hat{f} = -1.0t; F = 1.01t;\ \eta = 0.1;\ \gamma = 20;\ x_d = \sin(\pi t/2) \tag{97}$$

When the control law given by

$$u = t - (\pi^2/4)\sin(\pi t/2) - 20\dot{\tilde{x}} - (1.11t - \phi)\mathrm{sat}(s/\phi) \tag{98}$$

$$\dot{\phi} = -20\phi + 1.11t \tag{99}$$

is applied, the error as well as the boundary keep increasing. This phenomenon can be seen in Figure 4.19.

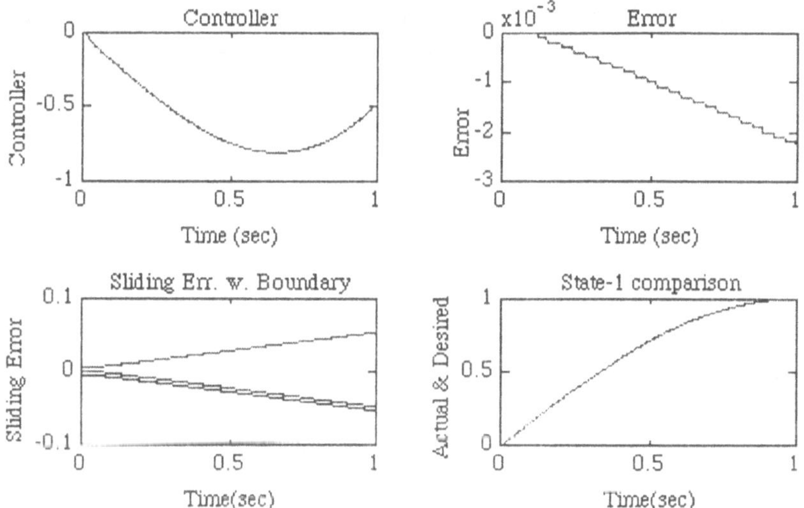


**Figure 4.19** *Simulation results on form-2 system using variable width sat function*

The chattering surface is shifted from 0 to the boundary ( $\phi$ = 0.02 here), when the control law given by

$$u = 1 - (\pi^2/4)\sin(\pi t/2) - 20\dot{\tilde{x}} - 1.11\text{msat}(0.0018,s,0.02) \quad (100)$$

is applied to the form-1 system with (94). This phenomenon is seen (Figure 4.20) because the trajectories inside the boundary as well as outside are attracted towards the lower boundary.

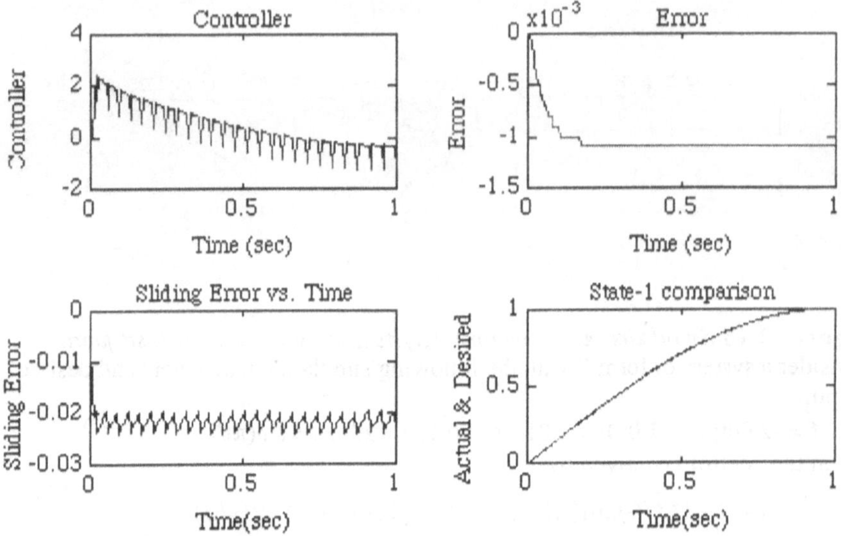


**Figure 4.20**: *Simulation results on form-1 system using msat function*

When the control law with the sat function and a fixed boundary is used for the system with (94), for a low value of sliding gain k, s tends to hit the boundary and stay there as shown in Figure 4.21. For a higher value of k it would either start chattering across the lower boundary or across the whole boundary width depending on the value.

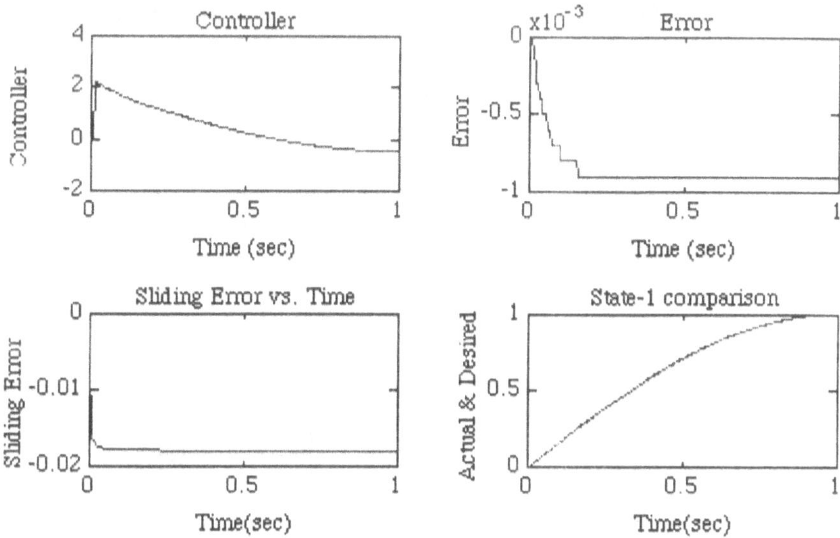


**Figure21**: *Simulation results on form-1 system using sat function*

Finally, the result of using the control law

$$u = 1 - (\pi^2/4)\sin(\pi t/2) - 20\dot{\tilde{x}} - (7.5 - \dot{\phi})\mathrm{int}(1,10,s,\phi) \tag{101}$$

with

$$\dot{\phi} = -20\phi + 7.5 \tag{102}$$

to the system (33) with form-1 parameters (94) is shown in Figure 4.22. Here, the output error and s go to zero. Instead of using a variable boundary with the int function, we could also have taken a fixed boundary.

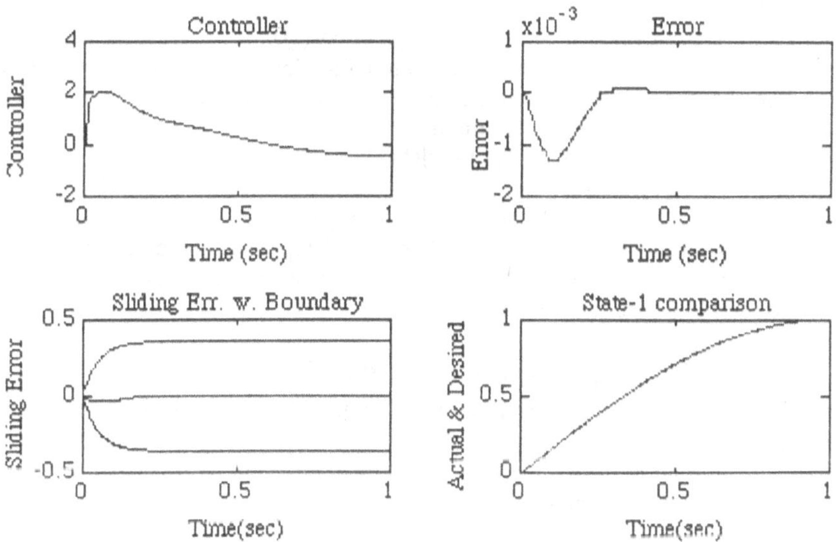


**Figure 4.22:** *Simulation results on form-1 system using int function*

# 6. Generalization of Chattering Reduction Results

The claim made in Section 4 can be generalized to a class of nonlinear systems by using the internal model principle approach [19]. Consider the nonlinear systems (33) and (42). Denote the input to the filters described in Section 4, for both nonlinear systems, by $d(\mathbf{x}_d,t)$. Note that $\mathbf{x}_d(t)$ is a function of time so that we can write the input as d(t). The filter for the general problem is shown in Figure 4.23, where $G_c(p)$ is to be designed by replacing the signum function in the boundary by some appropriate function to drive s to zero.

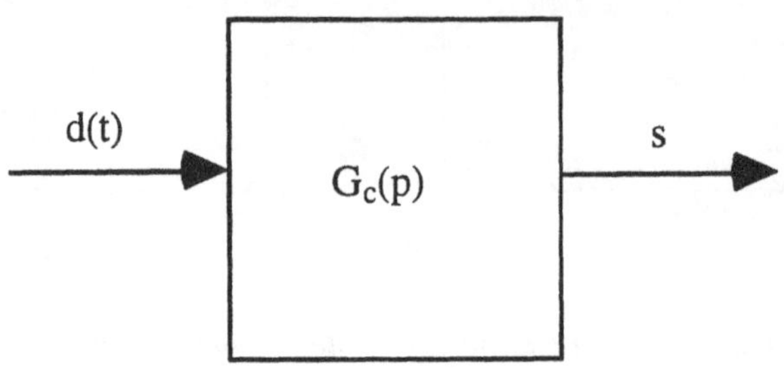


**Figure 4.23**: *A general filter*

Consider the class of nonlinear systems such that d(t) satisfies

$$A(p)d(t) = 0 \tag{103}$$

Some examples of such d(t)s are given below:

a. $pd(t) = 0$ for $d(t) =$ constant.

b. $p^2d(t) = 0$ for $d(t) = t$

c. $(p^2 + \omega^2)d(t) = 0$ for $d(t) = \sin(_t)$ or $\cos(_t)$.

Outside the boundary layer we use the signum function, but inside the boundary we change the control law by substituting the signum function by a modified function to obtain the following dynamics:

$$\dot{s} = -f(s) + d(t) \tag{104}$$

In this equation, f(s) is a function of s so that

$$L(f(s)) = [R(p) + T(p)/A(p)]S(p) \tag{105}$$

where L denotes the Laplace operator and, R(p) and T(p) are polynomials in p. S(p) is the Laplace transform of s(t). Taking the Laplace transform of Equation (104) and rearranging terms, we obtain

$$S(p) = [\frac{A(p)}{(p + R(p))A(p) + T(p)}]D(p) \tag{106}$$

In this equation, R(p) and T(p) can be chosen by using the Diophantine equation to place the poles in the left half plane of the complex variable p. In this way, the variable s will asymptotically go to zero. In Equation (106), D(p) is the Laplace transform of the input signal d(t).

Now, for the system (33), we can generalize the control law as

$$u = \hat{u} - k \text{ gen}(s) \tag{107}$$

where, the generalized function is given by

$$\text{gen}(s) = \text{sgn}(s) \quad \text{for } |s| \geq \phi$$
$$\text{gen}(s) = f(s)/k \quad \text{for } |s| < \phi \tag{108}$$

As an example, let us take a constant d(t). Obviously, we have A(p) = p. We can obtain the characteristic equation

$$p^2 + 2\gamma p + \gamma^2 = 0 \tag{109}$$

inside the boundary by choosing

$$R(p) = 2_$$
$$T(p) = \gamma^2 \tag{110}$$

The Laplace transform of the dynamic equation for s inside the boundary is

$$ps(p) = [-2\gamma s(p) - (\gamma^2/p)s(p)] + D(p) \tag{111}$$

which verifies control law (87).

# 7. Control Design for DTR Problem

In the previous sections we studied a class of nonlinear systems and how to design controllers which are robust to uncertainties. In the context of dynamic traffic routing with two alternate routes, the aim of the controller is to achieve and maintain equal travel times on the two routes. This is referred to in traffic assignment literature as a user-equilibrium condition. It is very difficult to estimate the travel time in the distributed setting, but as a first estimate, we can define the error variable as [20]:

$$e(t) = \int_0^{L_1} \rho_1 \, dx_1 - \int_0^{L_2} \rho_2 \, dx_2 \tag{112}$$

This error variable provides an indirect estimate of the difference between the travel times of the two alternate routes. It essentially represents the fact that if the numbers of vehicles in the two alternate routes are very different, then the travel time difference between the two will be large. Differentiating this equation with respect to time, we obtain

$$\dot{e}(t) = \int_0^{L_1} \frac{\partial \rho_1}{\partial t} dx_1 - \int_0^{L_2} \frac{\partial \rho_2}{\partial t} dx_2 \tag{113}$$

The terms inside the integral signs can be replaced by gradient terms using the conservation equation. Substituting Equations (4) and (5) in (113), yields

$$\dot{e}(t) = \int_0^{L_2} \frac{\partial q_2}{\partial x_2} dx_2 - \int_0^{L_1} \frac{\partial q_1}{\partial x_1} dx_1 \tag{114}$$

Equation (9) is same as

$$\dot{e}(t) = \int_{q_2(0)}^{q_2(L_2)} dq_2 - \int_{q_1(0)}^{q_1(L_1)} dq_1 \tag{115}$$

Performing the integration on the right hand side of the two terms, we get

$$\dot{e}(t) = [q_2(L_2) - q_1(L_1)] + [q_1(0) - q_2(0)] \tag{116}$$

We substitute the values of $q_1(0)$ and $q_2(0)$ from (6) into (116) and obtain

$$\dot{e}(t) = [q_2(L_2) - q_1(L_1)] + 2\beta U - U \tag{117}$$

Rearranging terms, we get,

$$\dot{e}(t) = [q_2(L_2) - q_1(L_1)] + 2\beta U - U \tag{118}$$

This can be written in the form:

$$\dot{x}(t) = f(x,t) + b(x,t)u(t) \tag{119}$$

where,

$$x = e \tag{120}$$

$$f = [q_2(L_2) - q_1(L_1) - U]$$

$$b = 2U$$

$$u = \beta$$

This form (119) is the same as equation (25) with n=1. Therefore, we can use the sliding mode control design technique presented in this chapter for this problem. We can also apply the chattering reduction techniques for that controller. The details of sliding mode control applicable to this problem are given in the references [18-20]. Choosing the sliding variable to be, i.e. s(t)=e(t) and using the notation and variables in [20], we obtain the control law:

$$u(t) = \hat{b}(x,t)^{-1}[\hat{u}(t) - k(x,t)\text{sign}(s(t))] \tag{121}$$

This law ensures that the travel time difference e(t) goes to zero in finite time. To implement this control law, equation (121) shows that we need to have the estimate of the flows at the ends of both sections and also the value of the error term. To measure the error term would involve having distributed sensing, such as a camera. If we use loops or other discrete sensors, then we can approximate the error term based on the sensor data.

A better estimate of the travel time can be obtained by integrating dx/v over the entire route length, where v is the velocity. Then, we obtain the difference of travel times between the two routes as:

$$e(t) = \int_0^{L_1} \frac{1}{v_1(\rho_1)} dx_1 - \int_0^{L_2} \frac{1}{v_2(\rho_2)} dx_2 \tag{122}$$

Differentiating this equation with respect to time and using chain rule, we obtain

$$\dot{e}(t) = \int_0^{L_1} \frac{\partial[1/v_1(\rho_1)]}{\partial\rho_1}\frac{\partial\rho_1}{\partial t} dx_1 - \int_0^{L_2} \frac{\partial[1/v_2(\rho_2)]}{\partial\rho_2}\frac{\partial\rho_2}{\partial t} dx_2 \tag{123}$$

Using (1) in (124) gives

$$\dot{e}(t) = \int_0^{q(L_1)} \frac{\partial[1/v_1(\rho_1)]}{\partial\rho_1} dq_1 - \int_0^{q(L_2)} \frac{\partial[1/v_2(\rho_2)]}{\partial\rho_2} dq_2 \tag{124}$$

We can substitute (3) here and perform the differentiation inside the integral sign to obtain

$$\dot{e}(t) = \int_0^{q(L_1)} \frac{\rho_{m_1}}{v_{f_1}(1-\frac{\rho}{\rho_{m_1}})^2} dq_1 - \int_0^{q(L_2)} \frac{\rho_{m_2}}{v_{f_2}(1-\frac{\rho}{\rho_{m_2}})^2} dq_2 \qquad (125)$$

From (2) and (3) we have $q = v_f \rho[1-(\rho/\rho_m)]$. Differentiating q with respect to ρ and substituting in (125) yields

$$\dot{e}(t) = \int_0^{\rho(L_1)} \frac{(\rho_{m_1}-2\rho_1)}{(\rho_{m_1}-\rho_1)^2} d\rho_1 - \int_0^{\rho(L_2)} \frac{(\rho_{m_2}-2\rho_2)}{(\rho_{m_2}-\rho_2)^2} d\rho_2 \qquad (126)$$

which is the same as

$$\dot{e}(t) = \int_0^{\rho(L_1)} \frac{2(\rho_{m_1}-\rho_1)}{(\rho_{m_1}-\rho_1)^2} d\rho_1 - \int_0^{\rho(L_1)} \frac{\rho_{m_1}}{(\rho_{m_1}-\rho_1)^2} d\rho_1 \qquad (127)$$

$$- \int_0^{\rho(L_2)} \frac{2(\rho_{m_2}-\rho_2)}{(\rho_{m_2}-\rho_2)^2} d\rho_2 + \int_0^{\rho(L_2)} \frac{\rho_{m_2}}{(\rho_{m_2}-\rho_2)^2} d\rho_2$$

Integrating (127), we get,

$$\dot{e}(t) = \ell n \frac{\rho_1(L_1)}{\rho_2(L_2)} + \frac{\rho_{m_1}}{[\rho_1(L_1)-\rho_{m_1}]} - \frac{\rho_{m_2}}{[\rho_2(L_2)-\rho_{m_2}]} \qquad (128)$$

$$+\ell n \frac{\rho_2(0)}{\rho_1(0)} - \frac{\rho_{m_1}}{[\rho_1(0)-\rho_{m_1}]} + \frac{\rho_{m_2}}{[\rho_2(0)-\rho_{m_2}]}$$

This can be written in the form (119) where

$$x = e \qquad (129)$$

$$f = \ell n \frac{\rho_1(L_1)}{\rho_2(L_2)} + \frac{\rho_{m_1}}{[\rho_1(L_1)-\rho_{m_1}]} - \frac{\rho_{m_2}}{[\rho_2(L_2)-\rho_{m_2}]}$$

$$b = 1$$

$$u = \ell n \frac{\rho_2(0)}{\rho_1(0)} - \frac{\rho_{m_1}}{[\rho_1(0)-\rho_{m_1}]} + \frac{\rho_{m_2}}{[\rho_2(0)-\rho_{m_2}]}$$

This form also can utilize the sliding mode control law (121) with the new parameters. Note that the relationship between the control variable for the sliding mode control law u and the actual control variable β is highly nonlinear. A lookup table can be used to obtain the value of β from u in real time.

# 8. Numerical Solution of Traffic PDE

There are many methods for obtaining numerical solutions to PDEs, such as finite difference method, weighted residual method, and Picard's iteration [21]. We will use the finite difference scheme in this chapter.

**Finite Difference Approximation**

Numerical solution of a pde can be obtained by using *finite difference approximation* derived from *Taylor series expansion* of functions. Finite difference scheme can be a forward difference scheme, a backward difference scheme, or a central difference scheme, as described below:

*Forward difference scheme*

If we choose two points x(i) and x(i+1), where x(i+1)=x(i)+h, then Taylor series expansion of a function f(x) gives us:

$$f(x(i+1)) = f(x(i)) + h\frac{df(x(i))}{dx} + O(h^2) \tag{130}$$

Dividing both sides of this equation and rearranging terms, we obtain:

$$\frac{df(x(i))}{dx} = \frac{f(x(i+1)) - f(x(i))}{h} + O(h) \tag{131}$$

This is a first order approximation of the first order derivative if we ignore second and higher order terms. This scheme is called forward difference scheme because the derivative depends on x(i+1) and x(i) (i.e. all the terms involved include x(i) or x(j) where j>i).

The term f(x(i+1))-f(x(i)) is called the first order forward *finite difference* and is symbolized as Δf(x(i)). We can obtain higher order finite differences by either using Taylor series expansion or the following:

$$\Delta^n f(x(i)) = \Delta^{n-1} f(x(i+1)) - \Delta^{n-1} f(x(i)) \tag{132}$$

$$\Delta^0 f(x(i)) = f(x(i))$$

A second order forward difference is give by:

$$\Delta^2 f(x(i)) = \Delta f(x(i+1)) - \Delta f(x(i)) \tag{133}$$

$$= f(x(i+2)) - 2f(x(i+1)) + f(x(i))$$

*Backward difference scheme*

If we choose two points x(i) and x(I-1), where x(I-1)=x(i)-h, then Taylor series expansion of a function f(x) gives us:

$$f(x(i-1)) = f(x(i)) - h\frac{df(x(i))}{dx} + O(h^2) \tag{134}$$

Dividing both sides of this equation and rearranging terms, we obtain:

$$\frac{df(x(i))}{dx} = \frac{f(x(i)) - f(x(i-1))}{h} + O(h) \tag{135}$$

This is also a first order approximation of the first order derivative if we ignore second and higher order terms. This scheme is called backward difference scheme because the derivative depends on x(i) and x(I-1) (i.e. all the terms involved include x(i) or x(j) where j<i).

The term f(x(i))-f(x(i-1)) is called the first order backward *finite difference* and is symbolized as $\Delta_b f(x(i))$. We can obtain higher order finite differences using Taylor series expansion or by using the following:

$$\Delta_b^n f(x(i)) = \Delta_b^{n-1} f(x(i)) - \Delta_b^{n-1} f(x(i-1)) \tag{136}$$

$$\Delta_b^0 f(x(i)) = f(x(i))$$

*Central difference scheme*

Central difference scheme utilizes terms less than and greater than i, such x(i-1) and x(i+1). Let us expand the following:

$$f(x(i+1)) = f(x(i)) + h\frac{df(x(i))}{dx} + \frac{h^2}{2}\frac{d^2f(x(i))}{dx^2} + O(h^3) \tag{137}$$

$$f(x(i-1)) = f(x(i)) - h\frac{df(x(i))}{dx} + \frac{h^2}{2}\frac{d^2f(x(i))}{dx^2} + O(h^3) \tag{138}$$

Subtracting (138) from (137), we get:

$$\frac{df(x(i))}{dx} = \frac{f(x(i+1)) - f(x(i-1))}{2h} + O(h^2) \tag{139}$$

Note that the central difference method provides better results since the approximation errors for this scheme are of second order as compared to first order for the other two schemes. The term f(x(i+1))-f(x(i-1)) is called the first order backward *finite difference* and is symbolized as $\delta(x(i))$. We can obtain higher order finite differences using Taylor series expansion or by using the following:

$$\delta^n f(x(i)) = \delta^{n-1} f(x(i)) - \delta^{n-1} f(x(i-1)) \tag{140}$$

$$\delta^0 f(x(i)) = f(x(i))$$

When a pde involves partial derivatives, then we treat those in the same way by keeping all other variables constant except the one with respect to which the dependent variable is being differentiated.

In order to solve a given pde using the finite difference scheme, we do the following tasks:

(4) Select a finite difference scheme
(5) Replace the appropriate terms in the pde by the finite difference terms
(6) Solve the resulting set of equations with the given boundary conditions

The finite difference scheme we use is the forward difference scheme and is shown in the next section with error analysis to determine the appropriate step sizes for convergence of the solution.

## 9. Error Analysis

The objective of this analysis is to gain a handle on the magnitude of the time step size $\Delta t$, as it relates to the distance step size $\Delta x$ and other system variables for each alternate highway. From Equations (1-3) one obtains Equation 24.

$$\frac{\partial}{\partial t}(\rho) = -\frac{\partial}{\partial x}\left\{\rho\ v_f - \frac{\rho^2}{\rho_{max}}\ v_f\right\} \tag{141}$$

Equation (141) in turn simplifies to Equation (142).

$$\frac{\partial}{\partial t}(\rho) = -\ v_f\left(1 - \frac{2\,\rho}{\rho_{max}}\right)\frac{\partial}{\partial x}(\rho) \tag{142}$$

Using the Euler approximation for the partial derivatives we get:

$$\frac{\rho(x_i,t_j+\Delta t)-\rho(x_i,t_j)}{\Delta t} - \frac{\partial^2}{\partial t^2}\rho(x_i,t_j+\xi_j\Delta t)\bullet \Delta t/2 = \tag{143}$$

$$-v_f\left(1-\frac{2\rho(x_i,t_j)}{\rho_{max}}\right)\{\frac{\rho(x_i+\Delta x,t_j)-\rho(x_i,t_j)}{\Delta x}$$

$$-\frac{\partial^2}{\partial t^2}\rho(x_i+\zeta i\Delta x,t_j)\bullet \Delta x/2\}$$

where $\Delta t$ is the time step size, $\Delta x$ is the distance step size, $x_i$ refers to the $i_{th}$ iteration in distance, $t_j$ refers to the $j_{th}$ iteration in time, and $0 < x_j, z_i < 1$.

Solving for the time update one gets Equation (144).

$$\rho(x_i,t_j+\Delta t) = \Delta t[\frac{\rho(x_i,t_j)}{\Delta t} \tag{144}$$

$$-v_f\left(1-\frac{2\rho(x_i,t_j)}{\rho_{max}}\right)\frac{\rho(x_i+\Delta x,t_j)-\rho(x_i,t_j)}{\Delta x}]+$$

$$+\Delta t[\frac{\partial^2}{\partial t^2}\rho(x_i,t_j+\xi_j\Delta t)\bullet \Delta t/2$$

$$-v_f\left(1-\frac{2\rho(x_i,t_j)}{\rho_{max}}\right)\left\{-\frac{\partial^2}{\partial t^2}\rho(x_i+\zeta i\Delta x,t_j)\bullet \Delta x/2\right\}]$$

We set $\alpha(x_i,t_j) = v_f(1-2\frac{\rho(x_i,t_j)}{\rho_{max}})$ (145)

The time update equation with error term is:

$$\rho(x_i,t_j+\Delta t)=\Delta t[\frac{\rho(x_i,t_j)}{\Delta t}$$
$$-\alpha(x_i,t_j)\frac{\rho(x_i,t_j)-\rho(x_i-\Delta x,t_j)}{\Delta x}]+$$
$$+\Delta t[\frac{\partial^2}{\partial t^2}\rho(x_i,t_j+\xi_j\Delta t)\bullet \Delta t/2$$
$$-\alpha(x_i,t_j)\left\{-\frac{\partial^2}{\partial t^2}\rho(x_i+\zeta i\Delta x,t_j)\bullet \Delta x/2\right\}] \tag{146}$$

Note that: $|\alpha(x_i,t_j)|\le v_f$ (147)

We set: $r=\alpha(x_i,t_j)\dfrac{\Delta t}{\Delta x}$ (148)

The time update equation becomes:

$$\rho(x_i,t_j+\Delta t)=\rho(x_i,t_j)-r\,\rho(x_i,t_j)+r\,\rho(x_{i-1},t_j)+$$
$$+\Delta t[\frac{\partial^2}{\partial t^2}\rho(x_i,t_j+\xi_j\Delta t)\bullet \Delta t/2$$
$$-\alpha(x_i,t_j)\left\{-\frac{\partial^2}{\partial t^2}\rho(x_i+\zeta i\Delta x,t_j)\bullet \Delta x/2\right\}] \tag{149}$$

To examine convergence we define the local truncation errors $e_{ij}=r_{ij}-r_{ij}^{*}$, where $r^{*}$ denotes the approximation of r.

Assuming sufficient differentiability of r, we see that for the closed subregion of W bounded by t <= T, that the error term (in brackets) is bounded by $\Delta t\,[\,M_1\,\Delta t + M_2\,\Delta x\,]$. For each j, let $E_j=\max|e_{ij}|$, and suppose that r <= 1. Then arguing recursively on (143) we have:

$$E_{j+1}\le E_j+\Delta t\,[M_1\,\Delta t\;+\;M_2\,\Delta x]$$
$$\le E_{j-1}+2\Delta t\,[M_1\,\Delta t\;+\;M_2\,\Delta x]\le\ldots\le$$
$$\le\;E_0+(j+1)\Delta t\,[M_1\,\Delta t\;+\;M_2\,\Delta x] \tag{150}$$

Letting $(\Delta x,\ \Delta t)\rightarrow(0,0)$, we have $E_j\rightarrow 0$ or $e_{ij}\rightarrow 0$, and hence it shows that the method converges for r<1.

From the time update equation, we obtain

$$e(x_i,t_j+\Delta t)=e(x_i,t_j)-r\,e(x_i,t_j)+r\,e(x_{i-1},t_j)$$
$$=(1-r)\,e(x_i,t_j)+r\,e(x_{i-1},t_j) \tag{151}$$

We have to have $0 \leq r \leq 1$, otherwise, if r < 0, then $E_{j+1} \leq (1-2r)E_j + \ldots$, and if r>1, then $E_{j+1} \leq (2r-1)E_j + \ldots$. Therefore, we can not prove convergence if we do not satisfy the constraint that r<1. Since $0 \leq r \leq 1$ and $|\alpha(x_i, t_j)| \leq v_f$ we infer that $v_f \frac{\Delta t}{\Delta x} \leq 1$ and consequently $\Delta t \leq \frac{\Delta x}{v_f}$ in order for the algorithm to converge.

**Boundary Conditions**

In general, when we apply finite difference scheme to a pde similar to the traffic model, we are given boundary conditions at both ends. The reason is that, at any point since the dynamics are given in terms of gradient of traffic flow (q), the behavior at any fixed x is dependent on behavior at x greater than and less than x. That is why, we see shock wave phenomenon travelling in both directions. When we use the forward difference scheme, we don't need the boundary condition at the terminal end. This essentially implies that there is actually a boundary condition for the original pde problem, but it is indirectly being applied.

# 10. Simulation Software

The software is developed in MATLAB environment. MATLAB is a very popular environment for mathematical simulations, and hence it was used for this study. We use two files for performing the simulation. File pde.m (listing 4.1) calls the function dtr_pde defined in the file dtr_pde.m (listing 4.2) and then draws various plots as shown in Figure 4.24. In order to run the software, the user has to type pde at the prompt on the command window as shown in Figure 4.25.

```
%      pde.m
clear;
clg;
clc;

[error,qin,beta,time,ro,ro2] = dtr_pde(10,2400);

subplot (221);
plot(time,error,'b');
xlabel('time');
ylabel('travel time difference');

subplot(222);
plot(time,beta,'b');
xlabel('time');
ylabel('split factor');

subplot (223);
mesh(ro);
```

```
xlabel('time');
ylabel('distance');
zlabel('traffic density for rout#1');

subplot(224);
mesh(ro2);
xlabel('time');
ylabel('distance');
zlabel('traffic density for rout#2');
```

**Listing 4.1**: *File pde.m*

```
function  [error,qin,beta,time,ro,ro2]=  dtr_pde(time_step,total_time)
%       dtr_pde.m
%       Matlab m-file  for using sliding mode control for
%       DTR problem in distributed parameter setting
%
%       dt and dx
%       time_step and total time are in seconds

vf=60;
ro_jam=120;

vf2=60;
ro_jam2=120;

dt=time_step/3600.0;  %1 second
dx=.1;                %      .1 mile

time_up = round(total_time/time_step);

%       Initialize Flow and Density Matrices at time t=0
%       The index is q(x,t) use L/100.
%       So the distance L=10 miles and each dx is .1 miles

for xi=1:101    % initial condition along the whole length
ro(xi,1)= 45;   % vehicles/mile
v(xi,1) = vf*(1-(ro(xi,1)/ro_jam));
q(xi,1)=ro(xi,1)*v(xi,1);

ro2(xi,1)= 10;  % vehicles/mile
v2(xi,1) = vf2*(1-(ro2(xi,1)/ro_jam2));
q2(xi,1)=ro2(xi,1)*v2(xi,1);
end;

%       Time update
for ti = 1:time_up
%if ti>(time_up/2) ro_jam2=119; end
%if ti>(time_up/2) vf2=59; end
clc;
```

```
ti
time(ti)=ti;

%      ESTIMATE THE FLOW AT TIME ti: ADD QO + RAMP FLOW
(INPUT)
%qin(ti) = 2500*(1+0.2*sin(ti/100));     %vehicles/mile
qin(ti) = 200;


%      Control Gain
K=0.0001;

W=0;          %      Performance measure
for i=0:100
W = W + (ro(i+1,ti)-ro2(i+1,ti));
end;
error(ti)=W;
%if abs(W) < 100 sat=W/1000;end
%if abs(W) >= 100 sat=sign(W);end
u=(qin(ti)+q(101,ti)-q2(101,ti)-K*W)/(2*qin(ti));
%if ti==1 uold=0; end
%u=0.5*ut+0.5*uold;
%if (ti==1|abs(W)>1000) u=ut; end
%uold=u;

%      control
beta(ti) = min(1,max(0,u));

qo_init = beta(ti)*qin(ti) ;
ro_init = ro_jam*(1-sqrt(1-(4*qo_init/(vf*ro_jam))))/2;

qo2_init = (1-beta(ti))*qin(ti) ;
ro2_init = ro_jam2*(1-sqrt(1-(4*qo2_init/(vf2*ro_jam2))))/2;

%      Update along the length of the road
ro(1,ti+1)=ro_init;      %      input
v(1,ti+1) = vf*(1-(ro(1,ti+1)/ro_jam));
q(1,ti+1)=ro(1,ti+1)*v(1,ti+1);

ro2(1,ti+1)=ro2_init;    %      input
v2(1,ti+1) = vf2*(1-(ro2(1,ti+1)/ro_jam2));
q2(1,ti+1)=ro2(1,ti+1)*v2(1,ti+1);


for xi=1:100
ro(xi+1,ti+1)=max(0,min(ro_jam,dt*((ro(xi+1,ti)/dt)+(q(xi,ti)/dx)-
(q(xi+1,ti)/dx))));
v(xi+1,ti+1) = vf*(1-(ro(xi+1,ti+1)/ro_jam));
q(xi+1,ti+1)=ro(xi+1,ti+1)*v(xi+1,ti+1);
```

```
ro2(xi+1,ti+1)=max(0,min(ro_jam2,dt*((ro2(xi+1,ti)/dt)+(q2(xi,ti)/dx)-
(q2(xi+1,ti)/dx))));
v2(xi+1,ti+1) = vf2*(1-(ro2(xi+1,ti+1)/ro_jam2));
q2(xi+1,ti+1)=ro2(xi+1,ti+1)*v2(xi+1,ti+1);
end;
```

**Listing 4.2**: *File dtr_pde.m*

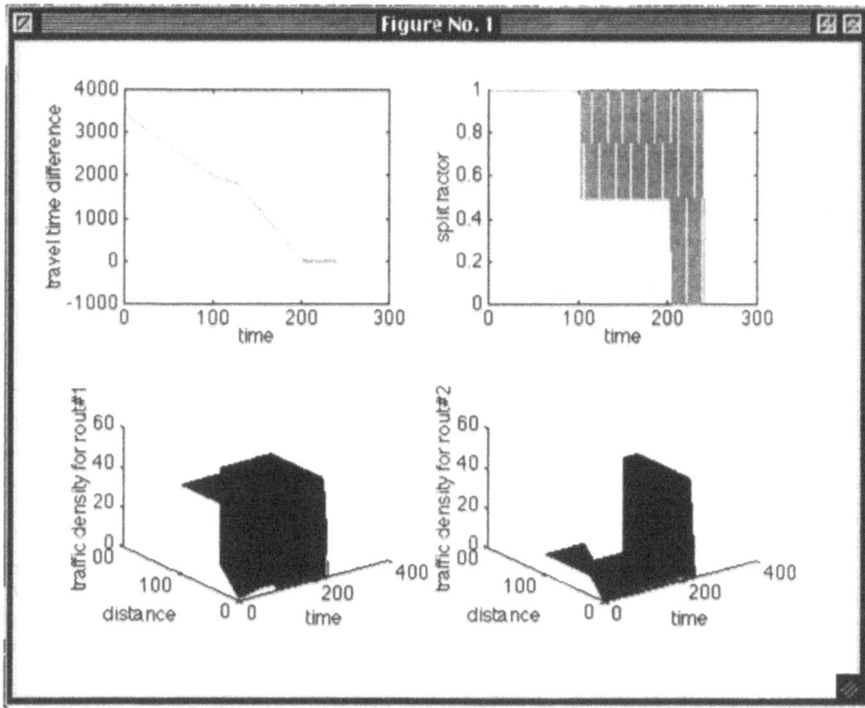


**Figure 4.24**: *Matlab Plots*

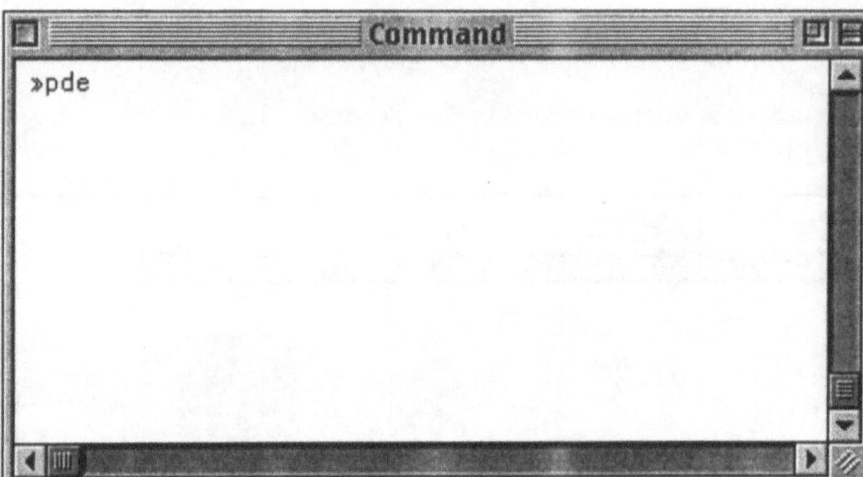


**Figure 4.25**: *Matlab Command Window*

## 11. Simulation Results

Simulations for assessing the performance of the controller were performed using Matlab environment. Figure 4.26 shows that the travel time difference between the two routes decreases to zero, as the sliding mode control is designed to achieve. The simulation also shows chattering at zero error, which also is a characteristic of this control. In actual implementation, the bandwidth of the control actuator will be limited by the human factors principles of not changing the commands given to human drivers that quickly. Hence, a filter would be used through which the control input variable would have to pass through.

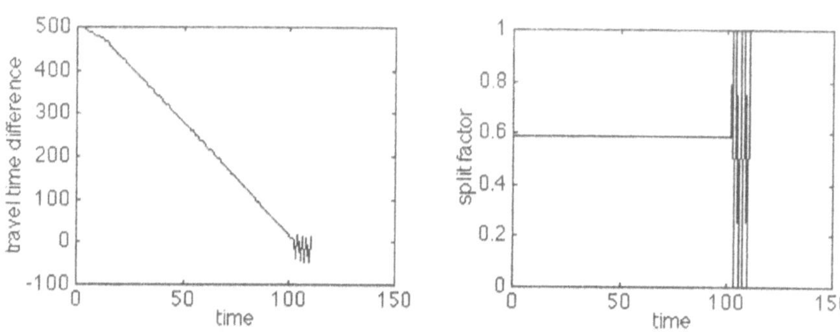


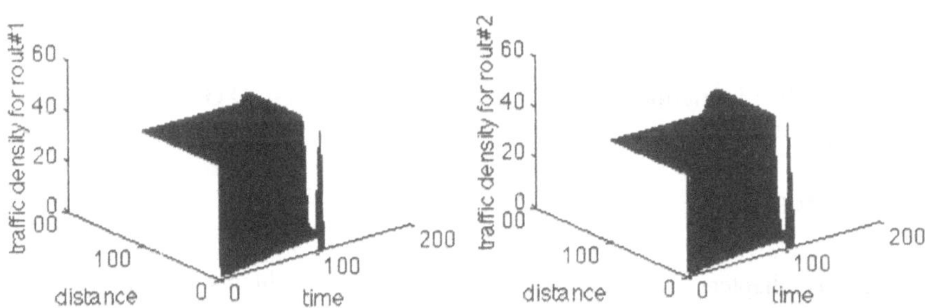


**Figure 4.26**: Simulation Results using Sliding Mode DTR control

Figure 4.27 shows that the travel time difference between the two routes also decreases to zero when we start with a much larger initial travel time difference. This simulation also shows chattering at zero error. The interesting result for this case is that, we see some chattering before the error reaches zero. This chattering is caused by the control action at boundary (x=0) affecting the other boundary (x=L) after the travel time in the second route. In general, the results of both the controllers show that the travel time difference in the alternate route is reduced to zero in finite time.

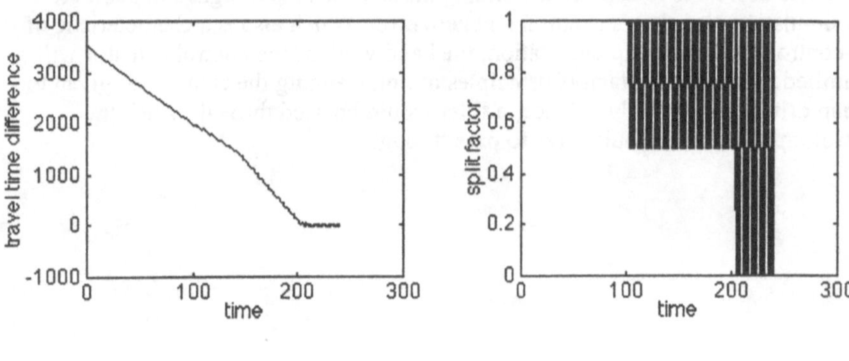


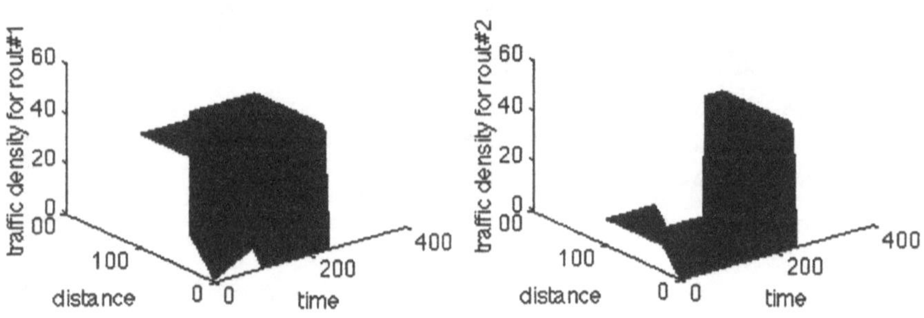


**Figure 4.27**: Simulation Results using Sliding Mode DTR control with a Relatively Large Initial Error

## 12. Summary

- In this chapter, we formulated feedback control problems in the distributed parameter setting.
- We designed a sliding mode feedback controller to achieve equal travel times on alternate routes using the control law.
- We used forward difference scheme to design a numerical simulation software for the problem.

## 13. Exercises

### Questions

Question 1: Why is sliding mode control appropriate for dynamic traffic routing problem of the kind studied in this chapter?
Question 2: How much time (finite or infinite time) does a trajectory take to reach zero state in a correctly designed sliding mode control system?
Question 3: What is a drawback of sliding mode control?
Question 4: What are the two ways chattering reduction can be performed for sliding mode control systems?
Question 5: What condition needs to be satisfied by the right hand side of the system of type (20) for a solution to exist?
Question 6: Is this condition satisfied by sliding mode control systems? If not, then how is a solution justified?
Question 7: What are different types of techniques for obtaining numerical solutions to PDEs?

### Problems

*Problem 1*

Write down the dynamic routing problem for two alternate routes (Figure 2) in terms of only the traffic densities on the two routes. Rewrite the boundary conditions also in terms of only the traffic densities. Show how this system could be solved by the method of characteristics if we were given the values of $\beta(t)$ and $U(t)$ for all t.

*Problem 2*

Consider the system

$$\dot{x}(t) = -\operatorname{sgn}(x(t))$$

If x(0)=2.5, then at what time will the system trajectory cross x=0? If x(0)=-2.5, then at what time will the system trajectory cross x=0?

*Problem 3*

Design a sliding mode control law to drive the following system to zero state.

$$\dot{x}(t) = 2.5x(t) + u$$

if there is a 20% uncertainty on the term 2.5x(t). Perform simulations of this system for various initial conditions. Design chattering reduction function for this controller.

*Problem 4*

Design a sliding mode control law to drive the following system to zero state.

$$\dot{x}(t) = 2.5x(t) + b(t)u$$

if there is a 20% uncertainty on the term 2.5x(t) and 1 < b(t) < 2. Perform simulations of this system for various initial conditions.

# 14. References

(1) M. Vidyasagar, Nonlinear Systems Analysis, Prentice Hall, Second Edition, 1992.

(2) A. F. Filippov, Differential equations with discontinuous right hand sides, *Mathematicheskii Sbornek*, 51, 1 (1960), in Russian. Translated in English, Am. *Math. Soc. Trans.*, 62, 199, 1964.

(3) A.F.Filippov, "Differential Equations with Second Members Discontinuous on Intersecting Surfaces," Differentsial'nye Uravneniya, vol.15, no.10, pp.1814-1832,1979.

(4) B. Paden and S. Sastry, "A Calculus for Computing Filippov's Differential Inclusion with Application to the Variable Structure Control of Robot Manipulators," IEEE Trans. Circ. Syst., vol. CAS-34, no.1, pp. 73-82, 1987.

(5) D. Shevitz, and B. Paden, "Lyapunov Stability Theory of Nonsmooth Systems,"EEE Trans. on Aut. Cont., vol.39, no.9, pp. 1910-1914, 1994.

(6) Pushkin Kachroo, "Existence of Solutions to a Class of Nonlinear Convergent Chattering Free Sliding Mode Control Systems," IEEE Transactions on Automatic Control, vol. 00, no. 00, September 1999.

(7) V. I. Utkin, *Sliding Modes and Their Application to Variable Structure Systems*, MIR Publishers, Moscow, 1978.

(8) Pushkin Kachroo, and Masayoshi Tomizuka, "Chattering Reduction and Error Convergence in the Sliding Mode Control of a Class of Nonlinear Systems", IEEE Transactions on Automatic Control, vol. 41, no. 7, July 1996.

(9) J. E. Slotine, and Weiping Li, *Applied Nonlinear Control*, Prentice Hall, New Jersey, 1991

(10) J. E. Slotine, and S. S. Sastry, Tracking Control of Nonlinear Systems Using Sliding Surfaces With Applications to Robot Manipulators, *Int. J. Control*, 39, 2, 1983.

(11) J. E. Slotine, and J. A. Coetsee, Adaptive Sliding Controller Synthesis for Nonlinear Systems, *Int. J. Control* , 1986.

(12) Bartolini and P. Pydynowski, "An improved chattering free VSC scheme for uncertain dynamical systems," IEEE Trans. Automat. Contr., vol. 41, pp. 1220-1226, 1996.

(13) G. Bartolini, A. Ferrara, and E. Usai , "Chattering Avoidance by Second-Order Sliding Mode Control," IEEE Transactions on Automatic Control, Volume 43 Number 2, February 1998.

(14) Bartolini and P. Pydynowski, "Asymptotic linearization of uncertain nonlinear systems by means of continuous control," Int. J. Robust Nonlin. Contr., vol. 3, pp. 87-103, 1993.

(15) Pushkin Kachroo, and Masayoshi Tomizuka, "Chattering Reduction and Error Convergence in the Sliding Mode Control of a Class of Nonlinear Systems", IEEE Transactions on Automatic Control, vol. 41, no. 7, July 1996.

(16) Pushkin Kachroo, "Nonlinear Control Strategies and Vehicle Traction Control.", Ph.D. dissertation at University of California at Berkeley, Department of Mechanical Engineering, 1993.

(17) Pushkin Kachroo, and Masayoshi Tomizuka, "Integral Action for Chattering Reduction and Error Convergence in Sliding Mode Control", American Control Conference, Chicago, 1992.

(18) Pushkin Kachroo, and Masayoshi Tomizuka, "Sliding Mode Control with Chattering Reduction and Error Convergence for a Class of Discrete Nonlinear Systems with Application to Vehicle Control", International Mechanical Engineering Congress & Expo., ASME 1995 Meeting, and ASME Journal of Dyn., Meas. & Control.

(19) B. A. Francis, and W. M. Wonham, The internal model principle of linear multivariable regulators, *Appl. Math. Optimiz.*, vol.2, no.2, pp. 170-194, 1975.

(20) Pushkin Kachroo and Kaan Özbay, "Sliding Mode for User Equilibrium Dynamic Traffic Routing Control", IEEE Conference on Intelligent Transportation Systems ITSC'97.

(21) J. F. Botha, and G. F. Pinder, *Fundamental Concepts in the Numerical Solution of Differential Equations*, John Wiley and Sons, 1983.

# CHAPTER 5
# DYNAMIC TRAFFIC ROUTING PROBLEM IN DISTRIBUTED PARAMETER SETTING USING SEMIGROUP THEORY

## Objectives

- To present the background information on semigroup theory and evolution equations
- To present the highway model in the semigroup context
- To give starting points for traffic routing control design using the semigroup theory

## 1. Introduction

Design of feedback controllers for systems that are modeled by partial differential equations can be performed using semigroup theory. In this technique, the system dynamics are written using operators, which make the dynamics look like ordinary differential equations in Banach space. The research work to design feedback controllers for traffic routing using this formulation is not complete and is in its infancy. This chapter merely presents the fundamentals of this approach and leaves it upto the reader to design the controllers.

The next section presents the basic concepts needed for the development, and the following section shows the system dynamics. That section is followed by some examples of how traffic system can be represented fort control design in this setting. Excellent references for the mathematical preliminaries are [1-4].

## 2. Mathematical Preliminaries

### Basic Topology

*Metric Space*

A metric space is a set X such that there is a real number d(p,q) called the **distance**, associated with any two elements p and q of the set with the following three properties

(1) $d(p,q) > 0 \quad \text{if } p \neq q; d(p,p) = 0;$

(2) *Symmetry* : $d(p,q)=d(q,p)$;

(3) *Triangle Inequality* : $d(p,q) \leq d(p,r) + d(r,p)$, for any $r \in X$

*Definitions*

The following definitions are stated with all the points and sets being elements and subsets of a given metric space X.

(1) **Neighborhood**: A neighborhood of radius r of a point p is a set of all points q, such that $d(p,q)<r$.
(2) **Limit Point**: A point p is a limit point of a set E if every neighborhood of p contains a point $q \neq p$, such that $q \in E$.
(3) **Closed Subset**: E is a closed subset of X if every limit point of E is also a point of E.
(4) **Interior Point**: A point p is an interior point of E if there exists a neighborhood N of p such that $N \subset E$.
(5) **Open Subset**: A subset E of X is open if every point of E is an interior point of E.
(6) **Compact Sets**: A subset K of a metric space X is called compact if every open cover of K contains a finite subcover.

**Vector Space**

*Definition*

A vector space (or a linear space) is a set L with two operations, multiplication and addition satisfying the following axioms.

(1) Given any $x \in L$ and $y \in L$ there exists a unique third element, $x + y \in L$, called the sum of x and y obtained through the **addition** operation, such that
    (1) $x+y = y+x, \ \forall x, y \in L$ *(Commutativity)*
    (2) $(x+y)+z = x+(y+z), \ \forall x, y, z \in L$ *(associativity)*
    (3) There exists a zero element $0 \in L$ such that $x+0 = x, \ \forall x \in L$ *(existence of additive identity)*
    (4) For each $x \in L$, there exists an element -x, such that $x+(-x) = 0$ *(existence of additive inverse)*
(2) Given any $\alpha \in R$ or C and $x \in L$ there exists a unique element, $\alpha x \in L$, called the product of $\alpha$ and x obtained through the **multiplication** operation, such that
    *(1)* $\alpha(\beta x) = (\alpha\beta)x$
    (2) $1x = x$
(3) The addition and multiplication operations follow two *distributive laws* :
    (1) $(\alpha + \beta)x = \alpha x + \beta x$
    (2) $\alpha(x + y) = \alpha x + \alpha y$

*Examples*

(1) The set of real numbers R is a linear space.

(2) The set of real vector $R^n$ is a linear space, where if $\mathbf{x} = (x_1, x_2, \ldots x_n)^T$ and $\mathbf{y} = (y_1, y_2, \ldots y_n)^T$, then addition is defined as $\mathbf{x} + \mathbf{y} = (x_1 + y_1, x_2 + y_2, \ldots x_n + y_n)^T$ and if $\alpha \in R$ or C, multiplication is defined as $\alpha\mathbf{x} = (\alpha x_1, \alpha x_2, \ldots \alpha x_n)^T$

(3) The set of complex vectors $C^n$ is a linear space.

(4) The set of real valued functions F[a,b] which map the compact interval [a,b] to R is a linear space, where for $x(.), y(.) \in F$,addition and multiplication are defined as $(x + y)(t) = x(t) + y(t), \forall t \in [a, b]$ and if $\alpha \in R$ or C, multiplication is defined as $(\alpha \cdot \mathbf{x})(t) = \alpha x(t), \forall t \in [a, b]$

(5) The set of real valued functions $F^n$ [a,b] which map the compact interval [a,b] to $R^n$ is a linear space, where for $\mathbf{x}(.), \mathbf{y}(.) \in F^n$,addition and multiplication are defined as $(\mathbf{x} + \mathbf{y})(t) = \mathbf{x}(t) + \mathbf{y}(t), \forall t \in [a, b]$ and if $\alpha \in R$ or C, multiplication is defined as $(\alpha \cdot \mathbf{x})(t) = \alpha\mathbf{x}(t), \forall t \in [a, b]$ where the addition and multiplication operations are defined in example (2) in this section.

*Subspace*

A subset S of a vector space L is a subspace if it satisfies the conditions:

(1) If $x,y \in S$, then $x+y \in S$.

(2) If $x \in S$ and $\alpha \in R$ or C, then $\alpha x \in S$.

**Normed Linear Space**

*Definition*

A linear space L equipped with a real-valued function $\|.\| : L \to R$ called the **norm** is called a normed linear space (L, $\|.\|$), if the norm function obeys the following axioms:

(1) $\|x\| \geq 0 \ \forall x \in L$, and $\|x\| = 0$ if and only if x=0.

(2) $\|\alpha x\| = |\alpha|.\|x\|, \ \forall \alpha \in R, \forall x \in L$

(3) *Triangle Inequality* : $\|x + y\| \leq \|x\| + \|y\| \ \forall x, y \in L$

*Examples*

Finite Dimensional

(1) The linear space of real valued vectors $R^n$ with the norm (called $\ell_\infty$ norm) $\|\mathbf{x}\|_\infty = \max_{1\le i\le n}|x_i|$ is a normed linear space.

(2) The linear space of real valued vectors $R^n$ with the norm (called $\ell_1$ norm) $\|\mathbf{x}\|_1 = \sum_{i=1}^{n}|x_i|$ is a normed linear space.

(3) The linear space of real valued vectors $R^n$ with the norm (called p norm) $\|\mathbf{x}\|_p = [\sum_{i=1}^{n}|x_i|^p]^{1/p}$ is a normed linear space, where $p\in[1,\infty]$. Examples (1) and (2) are special instances of this norm. When p=2, the norm is called the Euclidean or $\ell_2$ norm.

**Note:** All norms in $R^n$ are equivalent, i.e. for any vector $\mathbf{x}\in R^n$ and two norms $\|\cdot\|_a$ and $\|\cdot\|_b$, then there exist two constants $k_1$ and $k_2$ such that $k_1\|\mathbf{x}\|_a \le \|\mathbf{x}\|_b \le k_2\|\mathbf{x}\|_a$

Ininite Dimensional (Sequences)

(1) Let a sequence be $\{x_n\}_{n=1}^{\infty}$, $x_n \in R$. Let us define a norm $\|x\|_1 = \sum_{i=1}^{\infty}|x_i|$

Let us define a space $\ell_1$ of sequences as $\ell_1 = \{\{x_n\}_{n=1}^{\infty}, x_n \in R \mid \|x\|_1 < \infty\}$

The space $\ell_1$ is a normed linear space.

(2) Let a sequence be $\{x_n\}_{n=1}^{\infty}$, $x_n \in R$. Let us define a norm

$\|x\|_p = [\sum_{i=1}^{\infty}|x_i|^p]^{1/p}$ Let us define a space $\ell_p$ of sequences as

$\ell_p = \{\{x_n\}_{n=1}^{\infty}, x_n \in R \mid \|x\|_p < \infty\}$ The space $\ell_p$ is a normed linear space.

(3) Let a sequence be $\{x_n\}_{n=1}^{\infty}$, $x_n \in R$. Let us define a norm $\|x\|_\infty = \sup_{i\ge1}|x_i|$

Let us define a space $\ell_\infty$ of sequences as

$\ell_\infty = \{\{x_n\}_{n=1}^{\infty}, x_n \in R \mid \|x\|_\infty < \infty\}$ The space $\ell_\infty$ is a normed linear space.

**Note:** If x is a sequence and $x\in\ell_1$, then $x\in\ell_p$ and $x\in\ell_\infty$ for $p\in(1,\infty)$.

Ininite Dimensional (Functions)

(1) Let there be a given closed interval [a,b] in R and let C[a,b] denote the set of all continuous functions f:[a,b] $\rightarrow$R. Let us define a norm on this linear space as follows. If x(.)$\in$ C [a,b], then the norm of x(.), $\|x(.)\|_\infty$ is given as $\|x(.)\|_\infty = \max_{t\in[a,b]}|x(t)|$ The space (C[a,b], $\|x(.)\|_\infty$) is a normed linear space.

(2) Let there be a given closed interval [a,b] in R and let $C^n$[a,b] denote the set of all continuous functions f:[a,b] $\rightarrow$R. Let us define a norm on this linear space as follows. If $\mathbf{x}$(.)$\in$ $C^n$ [a,b], then the norm of $\mathbf{x}$, $\|\mathbf{x}(.)\|_\infty$ is given as $\|\mathbf{x}(.)\|_\infty = \max_{t\in[a,b]}\|\mathbf{x}(t)\|$ where $\|\mathbf{x}(t)\|$ is some vector norm on the vector $\mathbf{x}$(t). The space ($C^n$ [a,b], $\|\mathbf{x}(.)\|_\infty$) is a normed linear space.

(3) Let us consider functions $f(\cdot): R^+ \rightarrow R$. Let us define a norm

$\|f(\cdot)\|_p = [\int_0^\infty |f(t)|^p \, dt]^{1/p}, \quad p \in [1,\infty)$ Let us define a space $L_p$ of functions as $L_p = \{f(\cdot) \mid \|f(\cdot)\|_p < \infty\}$ The space $L_p$ is a normed linear space.

(4) Let us consider functions $f(\cdot): R^+ \rightarrow R$. Let us define a norm $\|f(\cdot)\|_\infty = \operatorname{ess\,sup}_{t\geq 0} |f(t)|$ Let us define a space $L_\infty$ of functions as $L_\infty = \{f(\cdot) \mid \|f(\cdot)\|_\infty < \infty\}$ The space $L_\infty$ is a normed linear space.

(5) Let us consider functions $\mathbf{f}(\cdot): R^+ \rightarrow R^n$. Let us define a norm

$\|\mathbf{f}(\cdot)\|_p = [\int_0^\infty \|\mathbf{f}(t)\|^p \, dt]^{1/p}, \quad p \in [1,\infty)$ where $\|\mathbf{f}(t)\|$ is any vector norm.

Let us define a space $L_p$ of functions as $L_p = \{\mathbf{f}(\cdot) \mid \|\mathbf{f}(\cdot)\|_p < \infty\}$ The space $L_p$ is a normed linear space.

(6) Let us consider functions $\mathbf{f}(\cdot): R^+ \rightarrow R^n$. Let us define a norm $\|\mathbf{f}(\cdot)\|_\infty = \operatorname{ess\,sup}_{t\geq 0} \|\mathbf{f}(t)\|$ Let us define a space $L_\infty$ of functions as $L_\infty = \{\mathbf{f}(\cdot) \mid \|\mathbf{f}(\cdot)\|_\infty < \infty\}$ where $\|\mathbf{f}(t)\|$ is any vector norm. The space $L_\infty$ is a normed linear space.

**Note:** For a function $f(\cdot): R^+ \rightarrow R$ if $f \subset L_1 \cap L_\infty$ then $f \in L_p, \quad p \in [1,\infty)$.

**Sequences**

*Convergent Sequence*

A sequence $\{p_n\}_{n=1}^\infty$ in a normed linear space (L, $\|.\|$) is said to *converge,* if there exists an element $p \in L$ with the property: for every $\varepsilon$>0, there exists an integer

N such that $n \geq N$ implies that $\|p_n - p\| < \varepsilon$. We denote this by $p_n \to p$ or $\lim_{n\to\infty} p_n = p$. The element p is called the **limit** of the sequence.

**Cauchy Sequence**

A sequence $\{p_n\}_{n=1}^{\infty}$ in a normed linear space (L, $\|.\|$) is said to be a *Cauchy sequence* if for every $\varepsilon$>0, there exists an integer N such that $n \geq N$, $m \geq N$ implies that $\|p_n - p_m\| < \varepsilon$.

**Banach Space**

*Definition*
A normed linear space (L, $\|.\|$) is called a Banach space or a **complete** normed linear space if every Cauchy sequence in (L, $\|.\|$) converges to an element of L.

*Examples*
(1) The space ($C^n$ [a,b], $\|\mathbf{x}(.)\|_{\infty}$) is a Banach space
(2) The space (C [a,b], $\|x(.)\|_{\infty}$) is a Banach space

**Inner Product Space**

*Definition*

A linear space L equipped with a real-valued or complex valued function $\langle\cdot,\cdot\rangle : L \times L \to F$ called the **inner product** is called an inner product space (L, $\langle\cdot,\cdot\rangle$), if the inner product function obeys the following axioms:
(1) $\langle x,x\rangle \geq 0 \ \forall x \in L$, and $\langle x,x\rangle = 0$ if and only if x=0.
(2) $\langle x,\alpha y\rangle = \alpha\langle x,y\rangle, \ \forall \alpha \in F, \forall x,y \in L$
(3) $\langle x,y\rangle = \langle y,x\rangle$, if $F = R, \langle x,y\rangle = \overline{\langle y,x\rangle}$, if $F = C, \forall x,y \in L$, where the over bar indicates complex conjugate
(4) $\langle x,y+z\rangle = \langle x,y\rangle + \langle x,z\rangle, \ \forall x,y,z \in L$

*Theorem*

With a norm defined on an inner product space (L, $\langle\cdot,\cdot\rangle$) as $\|x\| = \langle x,x\rangle^{1/2}$, (L, $\|.\|$) is a normed linear space.

**Hilbert Space**

Hilbert space is an inner product space that is complete in the norm defined by the inner product.

*Examples*

- The space $(R^n, \langle\cdot,\cdot\rangle)$ , where $\langle x, y\rangle = \sum_{i=1}^{n} x_i y_i$ is a Hilbert space.
- The space ($C^n$ [a,b], $\langle\cdot,\cdot\rangle_c$), where $\langle \mathbf{x}, \mathbf{y}\rangle_c = \int_a^b \langle \mathbf{x}(t), \mathbf{y}(t)\rangle dt$ is not a Hilbert space. An example, which shows that is a Fourier series of a discontinuous function converges to that discontinuous function. Hence, in that example a sequence of continuous functions converges to a discontinuous function that does not belong to the same space. We can make this space complete by adding the missing part to the space. The completed space is the space of Lebesgue measurable square integrable functions mapping [a,b] into $R^n$. Lebesgue measurable functions is a large class which covers continuous and discontinuous functions. Please read [2] carefully if you are not familiar with Lebesgue measure and integration.

**Sobolev Space**

Sobolev space $W^{m,P}(X), X \subset R^n$ is the Banach space of complex valued functions with all their generalized partial derivates upto order m (where the functions and their partial derivatives belong to $L^P$ space) [4]. The norm in this space is given by

$$\|f\|_{m,P} = [\sum_{0 \le k \le m} \|D^k f\|_P^P]^{1/P} \quad 1 \le P < \infty \tag{1}$$

$$\|f\|_{m,\infty} = \max\{\|D^k f\|_\infty, 0 \le k \le m\}$$

Generalized derivative is the same as the classical derivative if it exists, otherwise it allows for a larger class of functions such as function with disconinuous derivatives, See [4] for details.

**Semigroups**

Semigroup of operators is a family of operators T(t), t≥0 on X with the following properties.

- T(0)=I (Identity)
- T(t+s)=T(t)T(s)=T(s)T(t) $\forall$ t,s≥0.

Semigroups can be generated by operators which can be their infinitesimal generators, such as the linear operator $A_\varepsilon$ as shown in section 4. The semigroup operators operating on initial state produce the evolution of the system.

## 3. System Dynamics

In this chapter we use the Burgers' equation for the traffic model,

$$\frac{\partial}{\partial t}\rho(x,t)+\rho(x,t)\frac{\partial}{\partial x}\rho(x,t)=\varepsilon\frac{\partial^2}{\partial x^2}\rho(x,t) \tag{2}$$

with a solution obtained by taking the following limit:

$$\rho(x,t)=\lim_{\varepsilon\to 0}\rho^{\varepsilon}(x,t) \tag{3}$$

where $\rho^{\varepsilon}(x,t)$ satisfies (2) [5-10]. Using this form reduces the Burgers' equation formulation into the classical traffic model with no diffusion

Work on the feedback control of Burgers' equation has been performed [10-12] by some researchers in the past. Curtain [13] showed using Kielhofer's stability results for semi-linear evolution equations [14], that there exists a stabilizing feedback law which can be obtained from the linearized equation, when the domain of the output operator is a certain subspace of $L^2$ which contains the Sobolev space $H_0^1$, where for a given domain $\Omega$ with boundary $\partial\Omega$, $L^2(\Omega)$ is the space of all measurable functions f such that $\int_{\Omega}|f(x)|^2 dx<\infty$, and $H_0^1(\Omega)$ is the set of all functions f in $L^2(\Omega)$ such that the derivatives $f'$ (or $\nabla f$) are also in $L^2(\Omega)$ and $f|_{\partial\Omega}=0$, implying that f=0 on the boundary. Burns and Kang [6] show the design of Linear Quadratic Regulator (LQR) optimal controller for the linearized equation [12]. They also study a boundary control problem [7] which is relevant for the traffic control problem, since the control split factor enters the dynamics as a boundary injection.

## 4. Existing Work

If we define an operator $A_\varepsilon$ on $H=L^2(0,L)$ as

$$A_\varepsilon\varphi=\varepsilon\frac{\partial^2}{x^2}\varphi,\quad \forall\varphi\in D(A_\varepsilon)=H^2(0,L)\cap H_0^1(0,L) \tag{4}$$

the traffic flow Burgers' equation can be transformed into the following state space representation

$$\frac{d}{dt}\rho(t)=A_\varepsilon\rho(t)-f(t,\rho(t)),\quad \rho(0)=\rho_0\quad (t>0)$$
$$\text{where}\quad f(t,\rho(t))=-\rho(x,t)\frac{\partial}{\partial x}\rho(x,t) \tag{5}$$

on the Hilbert space $H=L^2(0,L)$, where $D(A_\varepsilon)$ is the domain of the operator $A_\varepsilon$ and $H^2(0,L)=\{f\in L^2(0,L)\mid f',f''\in L^2(0,L)\}$. The operator $A_\varepsilon$ generates S(t), the $C_0$ semigroup which is given by

$$S(t)\rho=\sum_{n=1}^{n=\infty} e^{\lambda_n t}\langle\rho,\phi_n\rangle\phi_n, \tag{6}$$

where, the eigenvalues $\lambda_n$, n = 1, 2..., which constitute the spectrum $\sigma(A_\varepsilon)$ of the operator $A_\varepsilon$, are given by

$$\lambda_n=-\varepsilon n^2\pi^2/L^2, \tag{7}$$

and for each eigenvalue $\lambda_n$, the corresponding eigenfunction $\phi_n$, is given by

$$\phi_n(x)=\sqrt{\frac{2}{L}}\sin\frac{n\pi L}{x}, \quad 0<x<L. \tag{8}$$

It can be shown that the $C_0$ semigroup S(t) satisfies the following stability property

$$\|S(t)\|_{L(H)}\le e^{-(\varepsilon\pi^2/L^2)t}, \quad t\ge 0 \tag{9}$$

where L(H) is the space of all bounded linear operators from H to itself with the norm $\|T\|_{L(H)}=\sup_{x\in H,x\neq 0}(\|T(x)\|_{(H)}/\|x\|_H)=\sup_{x\in H,\|x\|=1}(\|T(x)\|_{(H)})$ for any $T\in L(H)$. Since, a partial differential equation can be regarded as an evolution system, this semigroup framework can be applied to the treatment of the control problem governed by partial differential equations.

The control input in the Burgers' equation can come either through the equation itself through affine control or through the boundary. The solution of boundary problem is based on the one for affine control problem, hence both cases are presented here.
For the first case, the standard linear control form is

$$\frac{d}{dt}\rho(t)=A_\varepsilon\rho(t)+Bu(t), \quad \rho(0)=\rho_0, \tag{10}$$
$$y(t)=C\rho(t).$$

Here, B and C are input and output operators respectively. Let us define the following Hilbert spaces.

$$U=R, W=H_0^1(0,L), Y=R^{k+m} \tag{11}$$

The operators $B:U\to H$, and $C:W\to Y$ are defined by

$$Bu=b(.)u, \quad \text{and} \quad C\rho=(\tilde{z}(\tilde{x}_1),...\tilde{z}(\tilde{x}_k),\tilde{z}'(\tilde{y}_1),...\tilde{z}'(\tilde{y}_m)) \tag{12}$$

where $b(.)\in H, u\in U, \tilde{x}_i\in(0,L), 1\le i\le k$, $\tilde{y}_j\in(0,L), 1\le j\le m$, $\tilde{z}(\tilde{x}_i)$ and $\tilde{z}'(\tilde{y}_j)$ are defined by

$$\tilde{z}(\tilde{x}_i)=\frac{1}{2\delta}\int_{\tilde{x}_i-\delta}^{\tilde{x}_i+\delta} z(s)ds \quad \text{and} \quad \tilde{z}'(\tilde{y}_j)=\frac{1}{2\delta}\int_{\tilde{y}_i-\delta}^{\tilde{y}_i+\delta} z'(s)ds \tag{13}$$

The constraint on $\delta>0$ is such that $(\tilde{x}_i-\delta,\tilde{x}_i+\delta)\subset(0,L)$ and $(\tilde{y}_j-\delta,\tilde{y}_j+\delta)\subset(0,L)\ \forall 1\le i\le k, 1\le j\le m$. The solution to (A7) is given as

$$\rho(t)=S(t)\rho_0+\int_0^t S(t-s)Bu(s)ds,$$
$$y(t)=CS(t)\rho_0+C\int_0^t S(t-s)Bu(s)ds. \tag{14}$$

The LQR problem, which for any given $\alpha>0$ finds $\bar{u}(.)\in L^2(0,\infty;U)$ that minimizes the cost functional (15), has been solved by Burns and Kang [11, 12].

$$J(u)=\int_0^\infty [\|y(t)\|_Y^2+\|u(t)\|_U^2]e^{2\alpha t}dt. \tag{15}$$

The solution to this problem with constraints of stabilizability and detectability is given by

$$\bar{u}(t)=-e^{-\alpha t}B^*M\rho(t)e^{\alpha t} \tag{16}$$

where M, a unique nonnegative self-adjoint operator, satisfies the following algebraic Riccati equation:

$$(A_\varepsilon+\alpha I)^*M\rho+M(A_\varepsilon+\alpha I)^*\rho-MBB^*M\rho+C^*C\rho=0 \tag{17}$$

This feedback control is represented by the integral of the product of a functional gain and $\rho(s,t)$ so that the closed loop Burgers' equation is [6-7]

$$\frac{\partial}{\partial t}\rho(x,t)=\varepsilon\frac{\partial^2}{\partial x^2}\rho(x,t)-\rho(x,t)\frac{\partial}{\partial x}\rho(x,t)+b(x)\int_0^L k(s)\rho(s,t)ds$$
$$\rho(0,x)=\rho_0(x),$$
$$\rho(t,0)=\rho(t,L)=0 \tag{18}$$

where $k=-B^*M$.

The boundary control problem is stated as follows:

$$\frac{\partial}{\partial t}\rho(x,t)=\varepsilon\frac{\partial^2}{\partial x^2}\rho(x,t),\quad 0<x<L,\quad t>0,$$
$$\rho(0,x)=\rho_0(x),$$
$$\rho(t,0)=0,\rho(t,L)=u(t) \tag{19}$$

This can also be converted into a state space form by using the following Dirichlet map $\Gamma:R\to L^2(0,L)$ given by

$$\Gamma u=\frac{x}{L}u,\quad u\in R \tag{20}$$

This is so because $\frac{\partial^2[\Gamma u(t)]}{\partial x^2}=0$. The state space form is given by

$$\frac{d}{dt}\rho(t)=A_\varepsilon(\rho(t)-\Gamma u(t)),\quad \rho(0)=\rho_0 \tag{21}$$

Similar to the linear control problem, the solution of an LQR problem in this framework is obtained in terms of an algebraic Riccati equation. Note that for a nonlinear optimal control problem, the design will require solution of a Hamilton-Jacobi equation instead of a Riccati equation.

Note that the above treatment shows how to perform boundary injection control for a single highway, similar to the traffic ramp-metering problem. For the point diversion problem, we will obtain two different PDEs for the two routes with coupled boundary conditions. The measurement equations can be written to obtain the desired objective of same travel time. This, at present is still an open problem.

# 5. Summary

- This chapter presents very preliminary ideas on the use of semigroups and evolution equations to traffic routing problems.
- The method to perform traffic control on a single highway is shown but it needs to be developed to a level to apply it to routing networks.
- Semigroup theory allows PDE systems to be represented in a form which resembles ODEs.

# 6. Exercises

## Questions

Question 1: What is a metric space?
Question 2: Define vector space and give some examples.
Question 3: Define norm on a vector space.
Question 4: What is a normed linear space?
Question 5: What is a Cauchy sequence?
Question 6: What is a Banach space?
Question 7: What is a Hilbert space?
Question 8: What is a Sobolov space?

## Problems

### *Problem 1*

Take the case of two highways connected at both ends with traffic flowing from one end to the other. This is the same problem discussed in chapter 3 for the distributed parameter case. Write down the appropriate model for that problem.

*Problem 2*

Linearize the two PDEs for the two highways and show how a controller might be designed for this problem where the travel times have to be equated.

# 7. References


(1) M. Vidyasagar, *Nonlinear Systems Analysis*, Prentice Hall, 1993.

(2) Murray R. Spiegel, *Theory and Problems of Real Variables*, Schaum's Outline Series, Mcgraw-Hill, 1994.

(3) Aldo Belleni-Morante, *A Concise Guide to Semigroups and Evolution Equations*, World Scientific, 1994.

(4) Aldo Belleni-Morante, *Applied Semigroups and Evolution Equations*, Clarendon Press, Oxford, 1979.

(5) Cole, J. D., 'On a quasi-linear parabolic equation occuring in aerodynamics', Quart. Appl. Math. IX, 1951, 225-236.

(6) Glimm, J. and Lax, P., *Decay of Solutions of Systems of Nonlinear Hyperbolic Conservation Laws*, Amer. Math. Soc. Memoir 101, A.M.S., Providence, 1970.

(7) Hopf, E., 'The partial differential equation $u_t + uu_x = \mu u_{xx}$', Comm. Pure and Appl. Math 3, 1950, 201-230.

(8) Lax, P. D., *Hyperbolic Systems of Conservation Laws and the Mathematical Theory of Shock Waves*, CBMS-NSF Regional Conference Series in Applied Mathematics 11, SIAM, 1973.

(9) Maslov, V. P., 'On a new principle of superposition for optimization problems', *Uspekhi Mat. Nauk* 42, 1987, 39-48.

(10) Maslov, V. P., 'A new approach to generalized solutions of nonlinear systems', *Soviet Math. Dokl* 35, 1987, 29-33.

(11) Burns, J. A. and Kang S., 'A control problem for Burgers' equation with bounded input/output', Nonlinear Dynamics 2, 235-262, 1991.

(12) Burns, J. A. and Kang S., ' A stabilization problem for Burgers' equation with unbounded control and observation', Intl. Series of Num. Math., 100, 51-72, 1991.

(13) Curtain, R. F., 'Stability of semilinear evolution equations in Hilbert space', J. Math. Pures et Appl. 63, 1984, 121-128.

(14) Kielhofer, H., 'Stability and semilinear evolution equations in Hilbert space', Arch. Rat. Mech. Anal. 57, 1974, 150-165.

# CHAPTER 6
# FUZZY FEEDBACK CONTROL FOR DYNAMIC TRAFFIC ROUTING

## Objectives

- To review fuzzy logic fundamentals for control design.
- To design a fuzzy feedback control law for a sample DTR problem.

## 1. Introduction

Fuzzy logic can be used effectively to deal with uncertainty in decision-making processes. Fuzzy control is based on the fuzzy set theory proposed by Zadeh [1, 2, 3]. There are three major ways to design fuzzy controllers. In the first method, the controller tries to emulate a human-like control action by transforming linguistic terms into fuzzy variables [4-7]. The second method is to develop heuristic based fuzzy controllers. In the third method, the traffic network is represented as a fuzzy system and a control is designed by analyzing the fuzzy model. In this paper, we employ the second method.

Recently, fuzzy set theory has been applied to several challenging transportation problems. Lotan and Koutsopulos [8,9] presented a modeling framework for route choice in the presence of information based on the concepts from fuzzy set theory, approximate reasoning and fuzzy control. The proposed framework included models for driver's perception of network attributes, attractiveness of alternative routes, as well as models for reaction to information and the route choice mechanism itself. Sasaki and Akiyama [10] paper was to describe the judgement process of the human operator by using fuzzy logic and to develop a traffic control system for automatic on-ramp control of the expressway. The actual performance of the fuzzy control system was then studied on the Osaka-Sakai route.

Chen, May and Auslander [11] presented a fuzzy controller for freeway ramp metering on the San Francisco-Oakland Bay Bridge. The response of the freeway under different control schemes was tested using FRECON2 simulation program developed at the University of California at Berkeley. In order to compare the performance of the fuzzy controller with the performance of the existing ramp controller and an idealized controller, 2 scenarios that contain no incidents and eight scenarios that contain an incident of varying severity and location were employed. In general, the fuzzy controller rapidly and smoothly responded to the incidents by significantly reducing total delays caused by incidents.

The next section provides an overview of fuzzy logic and is adapted from [12] and [13]. Section following that provides an example fuzzy logic design for traffic routing control for point diversion.

## 2. Overview of Fuzzy Logic

An overview of fuzzy logic is provided with a comparison with the classical crisp logic.

### Crisp Sets

A crisp set is a collection of objects that are called members or elements of the set. We denote sets by uppercase letters and its members by lower case letters. To show that an object a is an element of a set A, we use the notation

$$a \in A \tag{1}$$

which can be read as "a belongs to A". We represent a does not belong to A by

$$a \notin A \tag{2}$$

A set can be described by listing all its members in braces separated by commas. This is called the *list method* (also called the *roster method*). The following is an example of this method:

$$A=\{1,2,3,4,5\} \tag{3}$$

Here, the set A has integers 1,2,3,4,5 as members. If the number of elements of a set is very large, this method is inconvenient, or if the set has infinite number of elements, then this method is unfeasible. In that case, we use the property method (also called the rule method). In this method, we describe the set in terms of some properties held by all its elements. In this method, we would represent (3) as

$$A = \{x \mid (x \in I) \text{ and } (0 < x < 6)\} \tag{3}$$

Here, I indicates the set of integers.

We can illustrate many concepts from crisp set theory using *Venn diagrams*. Usually Venn diagrams aonly shows relationships between sets and not anything about the members of sets. Diagrams which show the members of sets, and their relationships are called *Sagittal diagrams*. For example $a \in A$ can be represented as shown in Figure 6.1

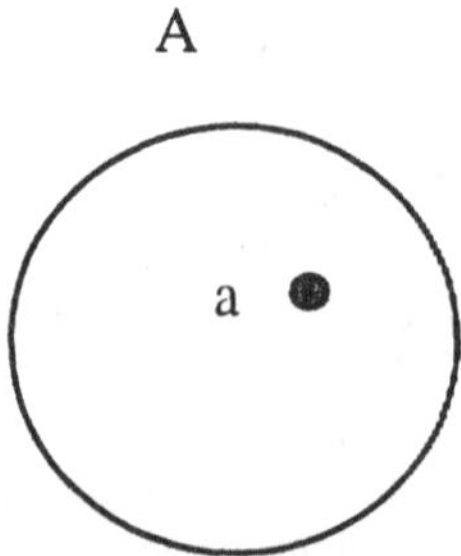


**Figure 6.1**: *Sagittal Diagram for* $a \in A$

*Subsets*

If every member of a set A is also a member of set B, i.e. if $x \in A$ implies $x \in B$, then A is called a subset of B and is represented as

$$A \subseteq B \tag{4}$$

Every set is a subset of itself. If $A \subseteq B$ and $B \subseteq A$ are true then A and B are called *equal sets* and are denoted as

$$A=B \tag{5}$$

To show that A and B are not equal sets, we use the representation

$$A \neq B \tag{6}$$

If $A \subseteq B$ but $A \neq B$, then we call A a proper subset of B and we represent this relationship as

$$A \subset B \tag{7}$$

This means that there is at least one member of B which is not a member of A. The following shows the relationship $A \subset B$ using a Venn diagram.

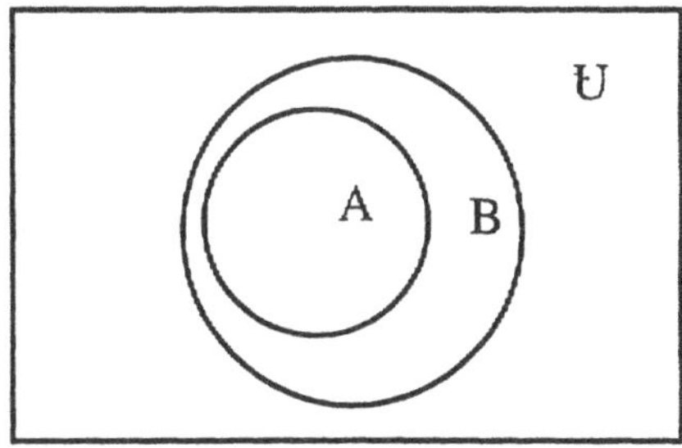


**Figure 6.2**: *Venn Diagram for* $A \subset B$

*Universal Set and Empty Set*

There exists a set containing all the elements. This set is called the universal set (or the *universe of discourse*) and is denoted by U. Every set is a subset of the universal set. There exists a set containing no elements. This set is called the empty set (also called null set) and is denoted by $\varnothing$. It is a subset of any set, and is a proper subset of any set except itself.

*Cardinality of Sets*

Cardinality of a set refers to the number of elements contained in the set. Cardinality can be finite or infinite. Cardinality of a set A is represented as $|A|$.

*Power Set*

Power set P(A) of a set A is a set containing all the subsets of A. Notice that $|P(A)| = 2^{|A|}$ is satisfied for any set A, which states that the cardinality of a power set of a set A is equal to 2 raise to the power of the cardinality of the set A.

*Membership Functions of Crisp Sets*

We use membership functions to determine if an object from the universal set is the member of a given set. For any crisp set A the membership function maps each member of the set with either a value 1 or 0, depending on whether x belongs to A or not. Precisely,

$$\mu_A : A \rightarrow \{0,1\} \tag{8}$$

And

$$\mu_A(x) = \begin{cases} 1 & \text{if } x \in A \\ 0 & \text{if } x \notin A \end{cases} \tag{9}$$

*Set Theoretic Operations*

Let A and B be two subsets of U. Then we define the following set operations:

Union: The union of two sets A and B is a set which contains all the elements which belong to A, or which belong to B. The union of A and B is represented as:

$$A \cup B = \{x \mid x \in A \quad \text{or} \quad x \in B\} \tag{10}$$

We can also describe the union operation in terms of membership function, as follows.

$$\mu_{A \cup B}(x) = \begin{cases} 1 & \text{if } x \in A \quad \text{or} \quad x \in B \\ 0 & \text{if } x \notin A \quad \text{and} \quad x \notin B \end{cases} \tag{11}$$

We can describe the membership function using a "max" function. However, this is not the only way to describe the union.

$$\mu_{A \cup B}(x) = \max[\mu_A(x), \mu_B(x)] \tag{12}$$

The union operation can be generalized for any number of sets as

$$\bigcup_{i \in I} A_i = \{x \mid x \in A_i \text{ for some } i \in I\} \tag{13}$$

The Venn diagram representation of $A \cup B$ is shown as the shaded portion in Figure 6.3.

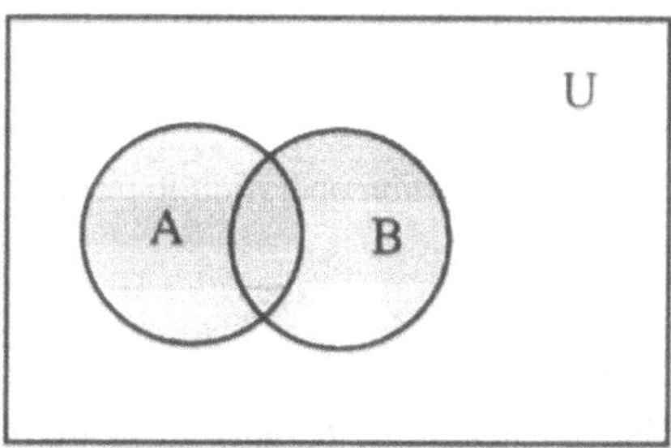


**Figure 6.3**: *Venn Diagram for* $A \cup B$

Intersection: The intersection of two sets A and B is a set which contains all the elements which are common to A and B. The intersection of A and B is represented as:

$$A \cap B = \{x \mid x \in A \quad \text{and} \quad x \in B\} \tag{14}$$

We can also describe the intersection operation in terms of membership function, as follows.

$$\mu_{A \cap B}(x) = \begin{cases} 1 & \text{if } x \in A \quad \text{and} \quad x \in B \\ 0 & \text{if } x \notin A \quad \text{or} \quad x \notin B \end{cases} \tag{15}$$

We can describe the membership function using a "min" function. However, this is not the only way to describe the intersection.

$$\mu_{A \cap B}(x) = \min[\mu_A(x), \mu_B(x)] \tag{16}$$

The intersection operation can be generalized for any number of sets as

$$\bigcap_{i \in I} A_i = \{x \mid x \in A_i \text{ for all } i \in I\} \tag{17}$$

The Venn diagram representation of $A \cap B$ is shown as the shaded portion in Figure 6.4.

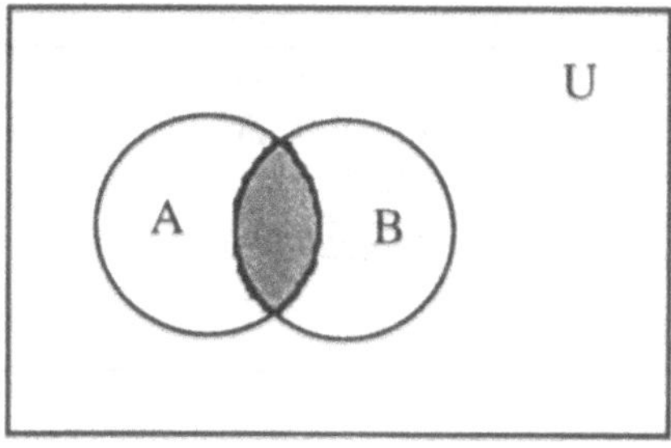


**Figure 6.4**: *Venn Diagram for* $A \cap B$

Complement: The complement of a sets A is a set which contains all the elements which are not in A. The complement of A is represented as $\overline{A}$ and is given by:

$$\overline{A} = \{x \mid x \notin A\} \tag{18}$$

We can also describe the intersection operation in terms of membership function, as follows.

$$\mu_{\overline{A}}(x) = \begin{cases} 1 & \text{if } x \notin A \\ 0 & \text{if } x \in A \end{cases} \tag{19}$$

We can describe the membership function of the complement of A in terms of the membership function of A as:

$$\mu_{\overline{A}}(x) = 1 - \mu_A(x) \tag{20}$$

The Venn diagram representation of $\overline{A}$ is shown as the shaded portion in Figure 6.5.

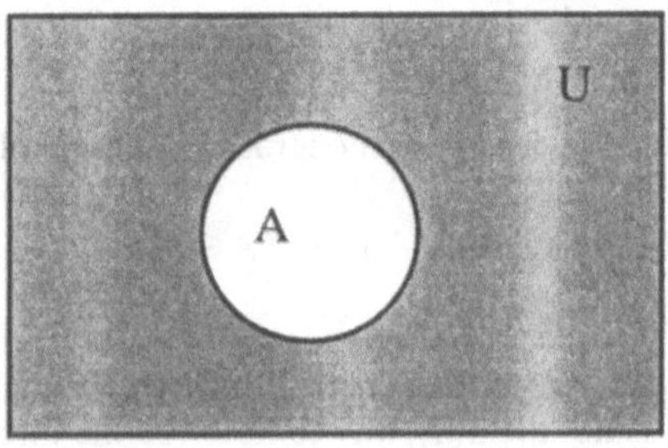


**Figure 6.5**: *Venn Diagram for* $\overline{A}$

*Selected Properties of Crisp Set Operations*

Some selected properties of crisp set operations are given in Table 6.1.

**Table 6.1** *Selected Properties of Crisp Set Operations*

|  |  |
|---|---|
| Idempotence | $A \cup A = A$ |
| | $A \cap A = A$ |
| Commutativity | $A \cup B = B \cup A$ |
| | $A \cap B = B \cap A$ |
| Law of Contradiction | $A \cap \overline{A} = \varnothing$ |
| Law of Excluded Middle | $A \cup \overline{A} = U$ |

*Crisp Relations*

Crisp relation shows existence or lack of existence of relationship between members of different sets. Let us illustrate this with two sets A and B and their members. Study the Figure 6.6, where the lines show existence of relationship between the members of the two sets A and B. Notice that, these relations can be one-to-many and many-to-one. Since, these can be one-to-many, these relations are not *functions* but just relations. The relation between sets A and B are represented by the notation R(A,B), as shown in Figure 6.6.

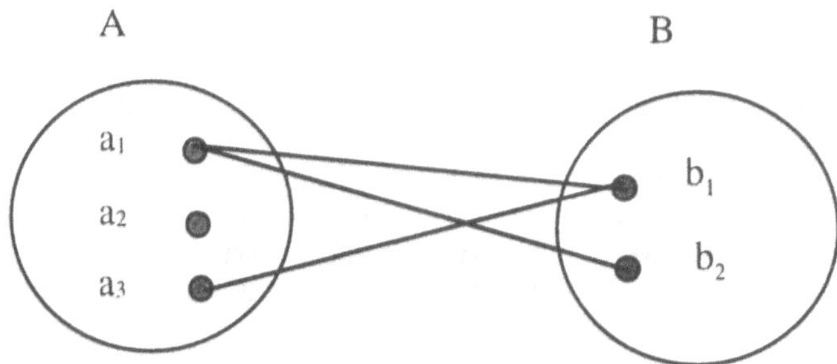


**Figure 6.6**: *Sagittal Diagram for* R(A,B)

The relationship of Figure 6.6 can be shown precisely with a matrix as shown in (21) below:

$$\begin{array}{c} \\ a_1 \\ a_2 \\ a_3 \end{array} \begin{array}{c} \begin{array}{cc} b_1 & b_2 \end{array} \\ \begin{pmatrix} 1 & 1 \\ 0 & 0 \\ 1 & 0 \end{pmatrix} \end{array} \tag{21}$$

The elements of the matrix take values of either 0 or 1, and they represent the membership of the element to the relation R(A,B). That is,

$$\mu_R(a,b) = \begin{cases} 1 & \Leftrightarrow (a,b) \in R(A,B), \text{i.e. a and b are connected} \\ 0 & \text{otherwise} \end{cases} \tag{22}$$

Notice that the set R(A,B) is a subset of the *Cartesian* product space $A \times B$, where

$$A \times B = \{(a,b) \mid a \in A \text{ and } b \in B\} \tag{23}$$

*Crisp Compositions*

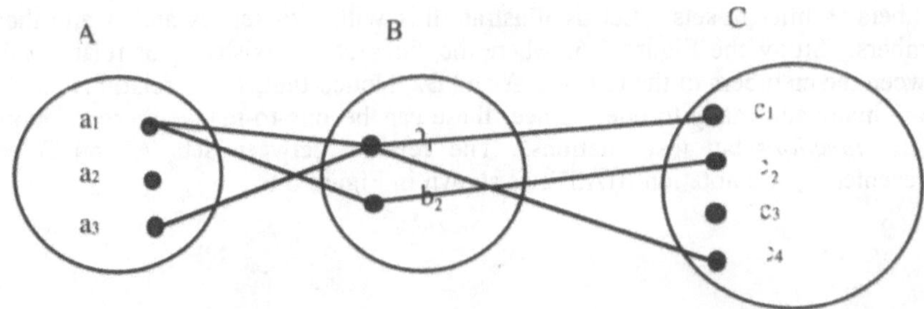


**Figure 6.7**: *Sagittal Diagram for* $R_1(A,B)$, $R_2(B,C)$, *and* $R_3(A,C)$

Compositions of relationships can be best understood by analyzing an example. We study the example of Figure 6.7. The composition of the relationships $R_1(A,B)$ and $R_2(B,C)$ where these relationships share a common space B, is another relationship $R_3(A,C)$. This composition $R_3(A,C)$ is represented by

$$R_3(A,C) = R_1(A,B) \circ R_2(B,C) \tag{24}$$

This is a subset of the product space $A \times C$, so that

$$(a,c) \in R_3(A,C) \Leftrightarrow \tag{25}$$

$$\exists b \in B, \text{s.t.}, (a,b) \in R_1(A,B) \text{ and } (b,c) \in R_2(B,C)$$

For the example of Figure 6.7, the sagittal diagrams for the three relationships are:

$$R_1(A,B) = \begin{matrix} & \begin{matrix} b_1 & b_2 \end{matrix} \\ \begin{matrix} a_1 \\ a_2 \\ a_3 \end{matrix} & \begin{pmatrix} 1 & 1 \\ 0 & 0 \\ 1 & 0 \end{pmatrix} \end{matrix}$$

$$R_2(B,C) = \begin{matrix} & \begin{matrix} c_1 & c_2 & c_3 & c_4 \end{matrix} \\ \begin{matrix} b_1 \\ b_2 \end{matrix} & \begin{pmatrix} 1 & 0 & 0 & 1 \\ 0 & 1 & 0 & 0 \end{pmatrix} \end{matrix}$$

$$R_3(A,C) = \begin{matrix} & \begin{matrix} c_1 & c_2 & c_3 & c_4 \end{matrix} \\ \begin{matrix} a_1 \\ a_2 \\ a_3 \end{matrix} & \begin{pmatrix} 1 & 1 & 0 & 1 \\ 0 & 0 & 0 & 0 \\ 1 & 0 & 0 & 1 \end{pmatrix} \end{matrix} \tag{26}$$

We can represent the composition of relationships in terms of membership functions as either a max-min composition or a max-product composition.

Max-min composition of relations $R_1(A,B)$ and $R_2(B,C)$ given by $R_3(A,C)$ and is defined in terms of the following membership function:

$$\mu_{R_1 \circ R_2}(a_i, c_k) = \left\{ (a_i, c_k), \max_{b_j} \left[ \min[\mu_{R_1}(a_i, b_j), \mu_{R_2}(b_j, c_k)] \right] \right. \quad (27)$$

We can check that (27) matches exactly with (26). For example, let us find out $\mu_{R_1 \circ R_2}(a_1, c_1)$ using (27). This is shown in Figure 6.8, where the solid line has membership 1 and dashed line has 0. We show dashed line instead of no line as in Figure 6.7 to show 0 membership. In this figure, we show only connections relating to $a_1$ and $c_1$.

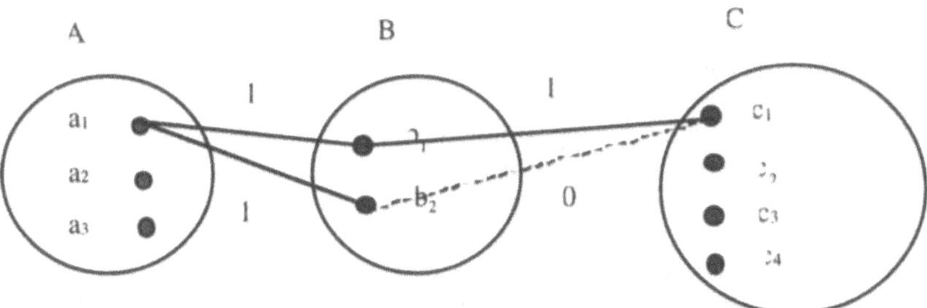


**Figure 6.8**: *Sagittal Diagram showing memberships for* $a_1$ *and* $c_1$

From this figure, we have

$$\min[\mu_{R_1}(a_1, b_1), \mu_{R_2}(b_1, c_1)] = 1 \quad (28)$$

and

$$\min[\mu_{R_1}(a_1, b_2), \mu_{R_2}(b_2, c_1)] = 0 \quad (29)$$

Therefore, from (28) and (29), we get,

$$\max[\min[\mu_{R_1}(a_1, b_1), \mu_{R_2}(b_1, c_1)]. \quad (30)$$

$$\min[\mu_{R_1}(a_1, b_2), \mu_{R_2}(b_2, c_1)]] = 1$$

which is the same value in (26) and is shown in Figure 6.9.

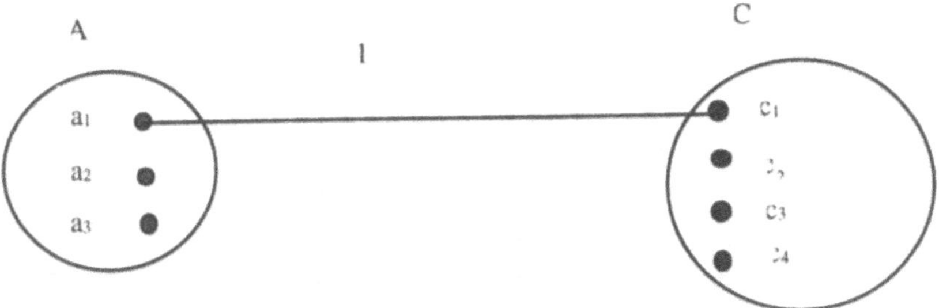


**Figure 6.9**: *Sagittal Diagram showing memberships for* $R_3(A,C)$ for $a_1$ *and* $c_1$

Another alternate formula to (27) is the max-product formula given by

$$\mu_{R_1 \times R_2}(a_i, c_k) = \left\{(a_i, c_k), \max_{b_j}\left[\mu_{R_1}(a_i, b_j), \mu_{R_2}(b_j, c_k)\right]\right\} \tag{31}$$

We can check that (31) also matches exactly with (26).

*Crisp Logic (Propositional Logic and Boolean Algebra)*

Logic is the formal study of reasoning and deduction. Propositional logic deals with manipulation of propositions (statements that can be either true or false). Proposition is a statement such as "Interstate I-81 is congested" which can either be true or false. This statement can denoted by a logic variable p which then can be manipulated.

p: Interstate I-81 is congested (32)

Similarly, there can be other propositions denoted by letters, q, r etc. We can define operations on two logic variables by using truth tables which shows the value of the operation for any possible combinations of the values of the logic variables. Since, each logic variable can have two values (i.e. true (1) or false (0)) there are four possible combinations of two logic variables p and q. For example, we can have p=0 and q=0, or p=0 and q=1, and so on. We can have sixteen possible functions for two variables since we have four digits for binary operation. The table below represents the three basic functions of the variables p and q from which the other functions can be created.

**Table 6.1**: *Truth Table for the Basic Logic Function*

| p<br>q | 1 1 0 0<br>1 0 1 0 | **Function Name** | **Representation** |
|---|---|---|---|
| | 1 0 0 0 | Conjunction | $p \wedge q$ |
| | 1 1 1 0 | Disjunction | $p \vee q$ |
| | 0 0 1 1 | Negation | p' |

Note that functions "conjunction", "disjunction" and "negation" are three of the possible sixteen functions possible. We can use these three primitive functions by combining them in algebraic expressions using the logic variables to produce logical formulas to represent the other functions. For instance, there is a function called "implication" and represented by $p \Rightarrow q$ (read as p implies q or as If p Then q, where p is called the *antecedent* and q the *consequent*). This function is described in Table 6.2.

**Table 6.2**: *Truth Table for implication*

| p<br>q | 1 1 0 0<br>1 0 1 0 | **Function Name** | **Representation** |
|---|---|---|---|
| | 1 0 1 1 | Implication | $p \Rightarrow q$ |

This function can be created from the primitives as follows:

$$(p \Rightarrow q) \Leftrightarrow (p' \vee q) \Leftrightarrow (p \wedge q')' \tag{33}$$

which states that $p \Rightarrow q$ is equivalent to $p' \vee q$ and is also equivalent to $(p \wedge q')'$. This equivalence is a *tautology*. Tautology is a statement that is always true. In this case, the truth table for the three statements is identically the same. The truth tables for the three statements are shown in Table 6.3.

**Table 6.3**: *Truth Table for implication and its equivalents*

| p | 1 1 0 0 | **Function Name** | **Representation** |
|---|---|---|---|
| q | 1 0 1 0 | | |
| | 1 0 1 1 | Implication | $p \Rightarrow q$ |
| | 1 0 1 1 | | $p' \vee q$ |
| | 1 0 1 1 | | $(p \wedge q')'$ |

The following tautology is very important in fuzzy logic control and is called *modus ponens*.

$$(p \wedge (p \Rightarrow q)) \Rightarrow q \tag{34}$$

Let us show that *modus ponens* is a tautology. We show it by using Table 6.4 where it is shown that this statement is true for any values of p and q.

**Table 6.4**: *Tautology of Modus Ponens*

| | | | | |
|---|---|---|---|---|
| p | 1 | 1 | 0 | 0 |
| q | 1 | 0 | 1 | 0 |
| $p \Rightarrow q$ | 1 | 0 | 1 | 1 |
| $p \wedge (p \Rightarrow q)$ | 1 | 0 | 0 | 0 |
| **Modus ponens $(p \wedge (p \Rightarrow q)) \Rightarrow q$** | **1** | **1** | **1** | **1** |

Boolean algebra deals with binary logic where the variables a,b,c, etc. can have values 0 or 1. Boolean algebra has operaions +, •, and ' which are analogous to 'or', 'and' and 'complement' operations in set theory and also 'disjunction', 'conjunction' and 'negation' in propositional logic. Properties of Boolean algebra is summarized in Table 6.5. In fact Boolean algebra, propositional logic and set theory are completely equivalent. Their equivalence is shown through the correspondence of some of their notations shown in Table 6.6. Table 6.5 and Table 6.1 provide some equivalent properties of set theory and Boolean algebra. Note that partial ordering in Boolean algebra can be defined as follows:

$$a \le b \Leftrightarrow a + b = b \tag{35}$$

or

$$a \le b \Leftrightarrow a \bullet b = a$$

Since, set theory, propositional logic and Boolean algebra are identical, we can consider only fuzzy sets rather than deal with fuzzy sets and fuzzy logic separately.

**Table 6.5**: *Some Properties of Boolean Algebra*

$a+a=a$

$a \bullet a=a$

$a+b=b+a$

$a \bullet b=b \bullet a$

$(a+b)+c=a+(b+c)$

$(a \bullet b) \bullet c=a \bullet (b \bullet c)$

a•(b+c)=(a•b)+(a•c)
a+(b•c)=(a+b) •(a+c)
a+(a•b)=a
a•(a+b)=a
a+a'=1
a•a'=0
1'=0
(a')'=a
(a+b)'=a'•b'
(a•b)'=a'+b'

**Table 6.6**: *Isomorphism between Set Theory, Propositional Logic and Boolean Algebra*

| Set Theory | Proposotional Logic | Boolean Algebra |
|---|---|---|
| U | 1 | 1 |
| ∅ | 0 | 0 |
| ∪ | ∨ | + |
| ∩ | ∧ | • |
| ¯ | ' | ' |
| ⊆ | ⇒ | ≤ |

**Fuzzy Sets**

Fuzzy sets are characterized by membership functions which take on values in the interval [0,1] unlike crisp sets which take values either 0 or 1. Precisely,

$$\mu_A : A \rightarrow [0,1] \tag{36}$$

Let us take an example to understand fuzzy sets. Let us say Jack is thirty years old. Is he young or old? The answer to that is not crisp. He can be considered old as well as young. We can give membership value to the age 30 in the set young and the set old. For example, we might have the following values $\mu_{old}(30) = 0.4$ and $\mu_{young}(30) = 0.8$. This states that the membership value for 30 years of age in the set "old" is 0.4, and in the set "young" is 0.8. Fuzzy set is a fuzzy subset of a universal set. For instance a set U of ages defined as U={10,20,30,40,50,60,70,80,90} has fuzzy sets 'young' and 'old'. Fuzzy sets are defined as a set of ordered pairs where the second element of each pair represents the membership value of the first element in the fuzzy set. A fuzzy set F then is represented as:

$$F = \{(x, \mu_F(x)) \mid x \in U\} \tag{37}$$

When U is discrete as in our example, then the fuzzy set F is shown as

$$F = \sum_U \mu_F(x)/x \tag{38}$$

It is very important to realize that the symbol / in (38) does not signify division and the summation using + does not indicate addition. The symbol / just shows the relationship between x and its membership value and the symbol + just shows the collection of all the finite points with their membership function. In our example the sets young and old could be

Young=1.0/10+0.9/20+0.8/30+0.4/40+0.3/50+0.1/60 (39)
Old=0.4/30+0.5/40+0.6/50+0.7/60+1.0/70+1.0/80+1.0/90 (40)

The membership functions and their graphical representation are shown in Figure 6.10.

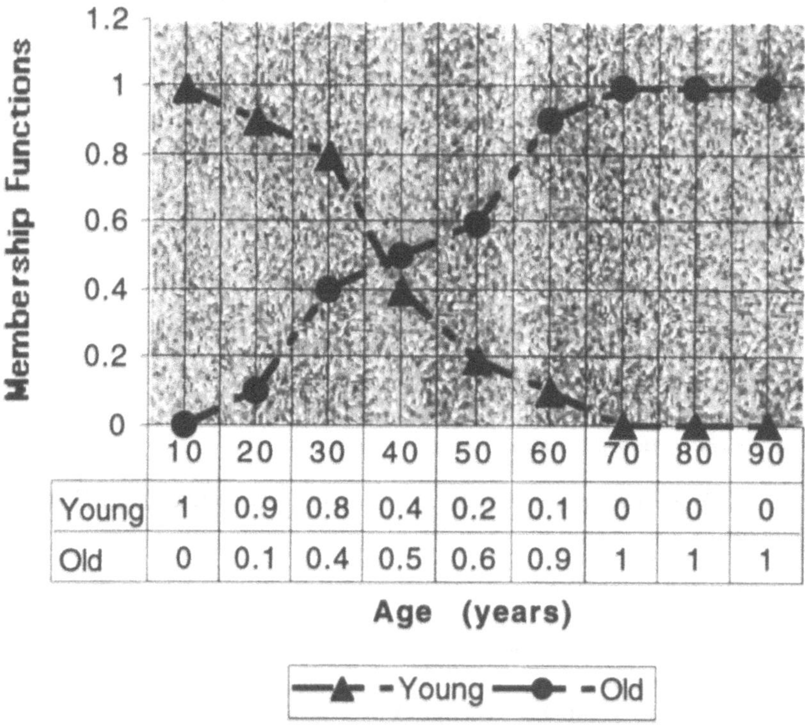


**Figure 6.10**: *Membership Functions for Young and Old.*

When U is continuous, then the fuzzy set F is shown as

$$F = \int_{U} \mu_F(x)/x \tag{41}$$

It is very important to realize in this case also that the symbol / in (41) does not signify division and the integral sign does not indicate integration. The symbol / just shows the relationship between x and its membership value and the integral symbol just shows the collection of all the infinite number of points with their membership function.

A variable like traffic flow is a continuous variable and would require fuzzy sets like (41). For instance, the variable traffic flow could have membership functions such as little, small, medium, large, and big. There are various types of membership functions proposed. The most common ones are triangular ones, as shown in Figure 6.11 for traffic flow. Other types of membership functions could be Gaussian, trapezoidal, etc.

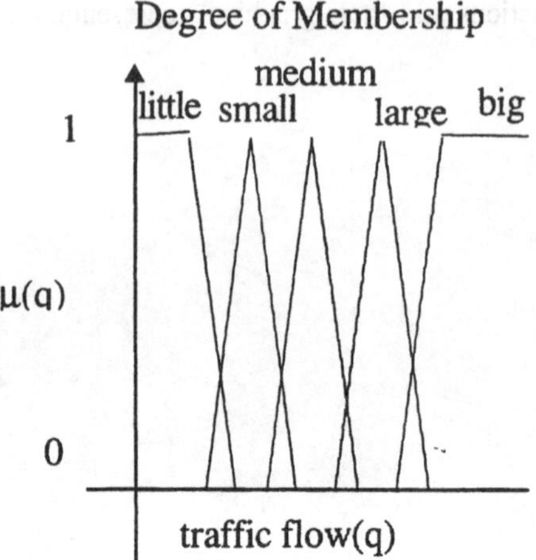


**Figure 6.11**: *Membership functions for Traffic Flow*

*Fuzzy Set Theoretic Operations*

Let A and B be two fuzzy subsets of U. Then we define the following set operations:

Fuzzy Union: The union of two fuzzy sets A and B is another fuzzy set which is defined in terms of a membership function obtained by using an operator on the membership functions of A and B. The operator for fuzzy union is called *t-conorm* (it is also called *s-norm*) and is denoted by ⊕. Hence, fuzzy union is obtained by performing the following operation.

$$\mu_{A \cup B}(x) = \mu_A(x) \oplus \mu_B(x) \tag{42}$$

There are many possible candidates for the t-conorm operator. The operator should satisfy some axioms for the fuzzy union, such as commutativity, associativity and monotinicity. It should also mimic crisp union for membership values of zero and one. One example of t-conorm is the Yager class of fuzzy unions given by:

$$\mu_A(x) \oplus \mu_B(x) = \min[1, ([\mu_A(x)]^w + [\mu_B(x)]^w)^{1/w}] \tag{43}$$

Here w can take on values between 0 and ∞. When we choose w= ∞, we obtain the following, which is one of the most often used fuzzy union operator.

$$\mu_{A \cup B}(x) = \max[\mu_A(x), \mu_B(x)] \tag{44}$$

Fuzzy union is shown graphically in Figure 6.12.

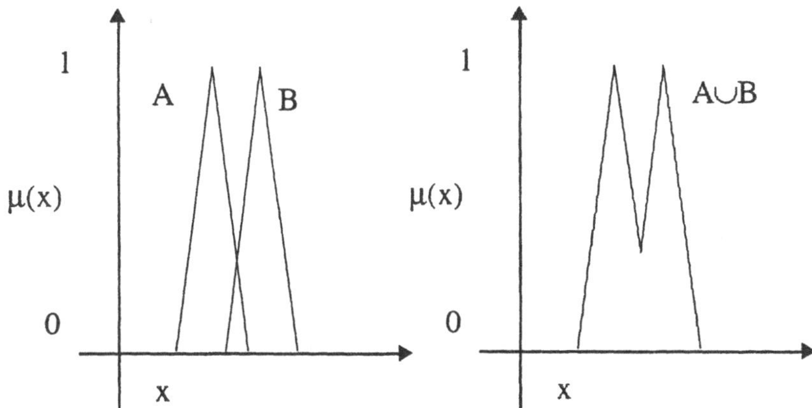


**Figure 6.12**: *Membership functions for Fuzzy Union*

Fuzzy Intersection: The intersection of two fuzzy sets A and B is another fuzzy set which is defined in terms of a membership function obtained by using an operator on the membership functions of A and B. The operator for fuzzy intersection is called *t-norm* and is denoted by *. Hence, fuzzy intersection is obtained by performing the following operation.

$$\mu_{A \cap B}(x) = \mu_A(x) * \mu_B(x) \tag{45}$$

There are many possible candidates for the t-norm operator. The operator should satisfy some axioms for the fuzzy intersection, such as commutativity, associativity and monotinicity. It should also mimic crisp intersection for membership values of zero and one. One example of t-norm is the Yager class of fuzzy intersections given by:

$$\mu_A(x) * \mu_B(x) = 1 - \min[1, ([1 - \mu_A(x)]^w + [1 - \mu_B(x)]^w)^{1/w}] \tag{46}$$

Here w can take on values between 0 and ∞. When we choose w= ∞, we obtain the following, which is one of the most often used fuzzy intersection operator.

$$\mu_{A \cap B}(x) = \min[\mu_A(x), \mu_B(x)] \tag{47}$$

Fuzzy intersection is shown graphically in Figure 6.13.

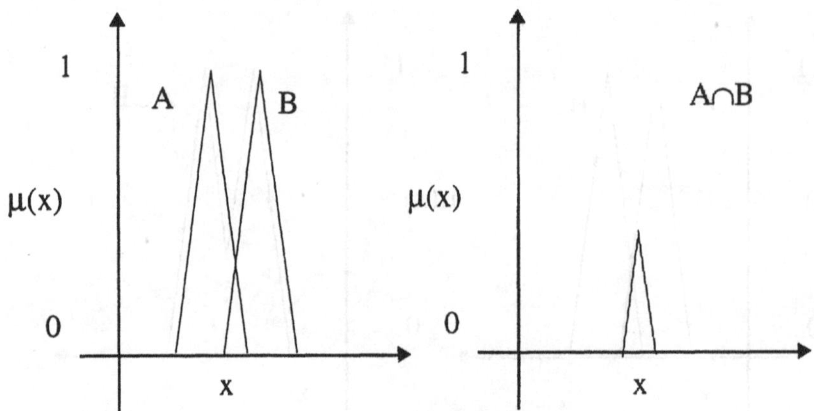


**Figure 6.13**: *Membership functions for Fuzzy Intersection*

Fuzzy Complement: The complement of a fuzzy set A is another fuzzy set which is defined in terms of a membership function obtained by using an operator on the membership function of A. The operator for fuzzy complement is denoted by c(.).Hence, fuzzy complement is obtained by performing the following operation.

$$\mu_{\bar{A}}(x) = c(\mu_A(x)) \tag{48}$$

There are many possible candidates for the complement operator. The operator should satisfy some axioms for the fuzzy intersection, such as monotinicity. It should also mimic crisp complement for membership values of zero and one. One example of complement is the Yager class of fuzzy complements given by:

$$\mu_{\bar{A}}(x) = [1 - \mu_A(x)]^{1/w} \tag{49}$$

When we choose w= 1, we obtain the following, which is one of the most often used fuzzy complement operator.

$$\mu_{\bar{A}}(x) = 1 - \mu_A(x) \tag{50}$$

Fuzzy complement is shown graphically in Figure 6.14.

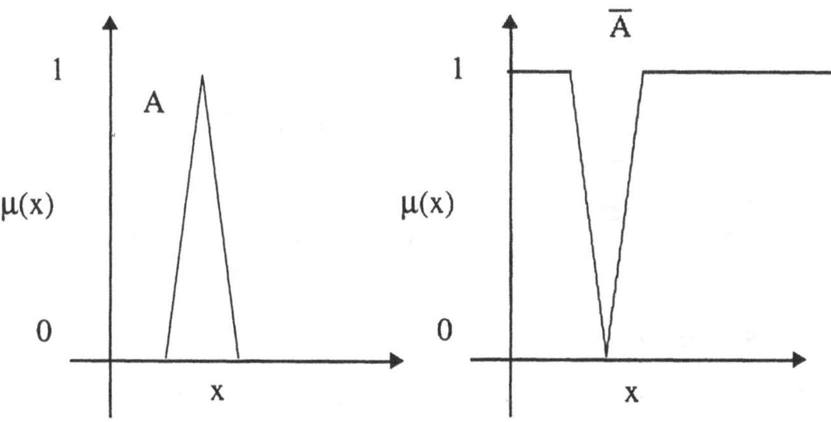


**Figure 6.14**: *Membership functions for Fuzzy Complement*

By definition of the fuzzy operators of union, intersection, and complement some of the counterparts of the crisp set operations for fuzzy sets are retained. However, there are examples of relationships, which are satisfied for crisp set operations but not fuzzy set operation. For example, the law of excluded middle and the law of contradiction, are not satisfied in fuzzy logic. This can be easily shown by using the triangular membership function of Figure 6.14.

*Fuzzy Relations and Compositions*

Fuzzy relation shows the degree of relationship between members of different sets. An example would be a statement "a is bigger than b", where a and b belong to different sets. In crisp set theory if a is 2 and b is 3, then the relationship between the two members is 0, however in fuzzy set theory we use membership functions to represent the relationship. Let us illustrate this with two sets A and B and their members. Study the Figure 6.15, where the numbers on the connecting lines show the degree of relationship between the members of the two sets A and B.

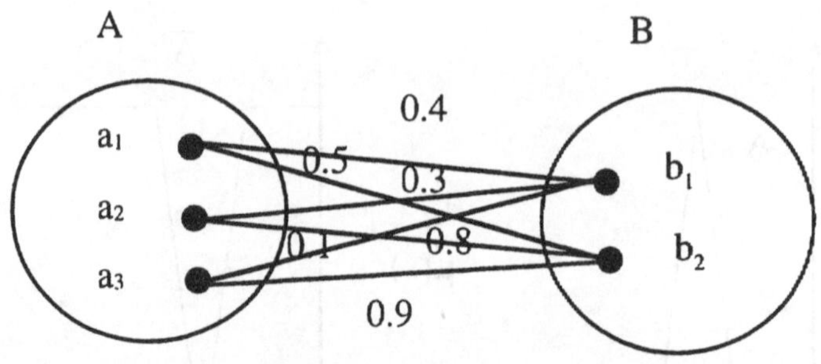


**Figure 6.15**: *Fuzzy Relation* R(A,B)

The relationship of Figure 6.6 can be shown precisely. with a matrix as shown in (21) below:

$$\begin{array}{c} \\ a_1 \\ a_2 \\ a_3 \end{array} \begin{array}{c} b_1 \quad b_2 \\ \begin{pmatrix} 0.4 & 0.5 \\ 0.3 & 0.8 \\ 0.1 & 0.9 \end{pmatrix} \end{array} \tag{51}$$

The elements of the matrix take values between 0 and 1, and they represent the membership of the element to the relation R(A,B). That is,

$$\mu_R(a,b) = w \quad \Leftrightarrow \text{a and b are connected with membership w} \tag{52}$$

Notice that the set R(A,B) is a subset of the *Cartesian* product space . Let $R_1(x,y)$ and $R_2(x,y)$ be two fuzzy relations on the same product space A×B, then we can define fuzzy union and intersections of these as:

$$\mu_{R_1 \cup R_2}(x,y) = \mu_{R_1}(x,y) \oplus \mu_{R_2}(x,y) \tag{53}$$

$$\mu_{R_1 \cap R_2}(x,y) = \mu_{R_1}(x,y) * \mu_{R_1}(x,y) \tag{54}$$

For composition of relationship from spaces A to B, and B to C in order to produce a relationship directly from A to C, we use the following *sup-star* composition of R and S which is motivated from (27) and (31).

$$\mu_{R_1 \circ R_2}(x,z) = \sup_{y \in B} \left[ \mu_{R_1}(x,y) * \mu_{R_2}(y,z) \right] \tag{55}$$

When A, B, and C are discrete, we can replace sup by max. In the case when the relationship $R_1$ is just a fuzzy set instead of a fuzzy relation (e.g. in the statement "x is high and if x is high then z is small"), then the composition becomes a function of only z as shown in (56).

$$\mu_{R_1 \circ R_2}(z) = \sup_{x \in A} \left[ \mu_{R_1}(x) * \mu_{R_2}(x,z) \right] \tag{56}$$

Sometimes in fuzzy control the rules might use linguistic modifiers on fuzzy variables. For example if we have a fuzzy set on low traffic, and if the rule uses the term very low traffic, we can produce a new fuzzy set low traffic by creating a new membership function by performing concentration (e.g. by reducing the width of a triangular function). Similarly some phrases might require dilation.

*Fuzzy Logic*

Fuzzy logic is used in terms of implication statements such as
"**If** Temperature is Low **AND** Humidity is High (57)
**Then** Output is Low"

This statement is of the form (A and B) implies C. We have dealt with "and" (intersection) operation in fuzzy logic using membership functions, but have not defined the operation for implication. We can calculate the membership function for p⇒q using the crisp relationship

$$(p \wedge (p \Rightarrow q)) \Rightarrow q \tag{58}$$

And hence, we could use

$$\mu_{p \Rightarrow q}(x,y) = \mu_{\bar{p} \cup q}(x,y) \tag{59}$$

Or

$$\mu_{p \Rightarrow q}(x,y) = c(\mu_{\bar{p} \cap q}(x,y)) \tag{60}$$

We can use (44) for fuzzy union, (47) for intersection and (50) for complement. However, using these relationships for implication does not produce membership functions, which are intuitive and practical. The membership functions of implications using (58) generally contain infinite support, which is undesirable. The following equations are the most popular ones in engineering for implications.

$$\mu_{p \to q}(x,y) = \min[\mu_p(x), \mu_q(y)] \tag{61}$$

$$\text{or } \mu_{p \to q}(x,y) = \mu_p(x)\mu_q(y)$$

*Fuzzy Logic Systems*

Since fuzzy logic systems interact with the crisp variables of the outside world, there is a need for transforming crisp variables to fuzzy variables, as well as to change fuzzy variables to crisp. This process is similar to the D/A and A/D conversion in digital-analog systems. Once, the input variables are converted to fuzzy variables, then these can be manipulated using fuzzy calculus, and then the fuzzy output can be converted to a crisp value which is then sent out. This is again similar to a microprocessor system, which first changes analog values to digital values using D/A conversion, then performs all the manipulation in digital domain, and then the output is converted to analog values. This overall framework is shown in Figure 6.16 below.

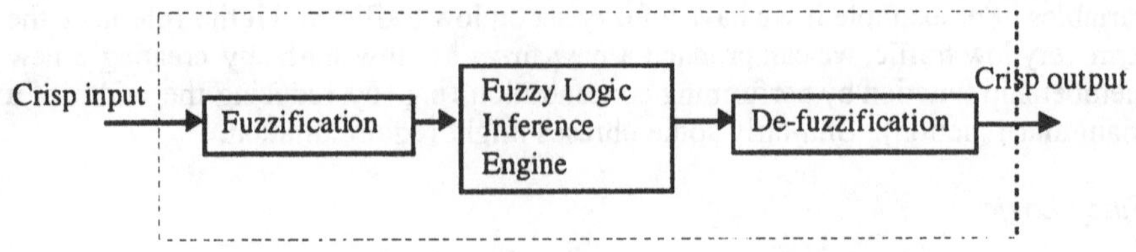


**Figure 6.16**: *Fuzzy Logic System*

Let us understand fuzzy logic system with an example. Let us consider a fuzzy logic system that has two inputs from sensors, which give the measurements of temperature and humidity. The output of the fuzzy logic system is a thermostat controller. There are two distinct phases of the fuzzy logic system: design and implementation.

Design Phase: During the design stage, the designer decides on the following:

4. what kind of membership functions will be used for each variable and how many fuzzy sets will be used for each variable, and
5. which fuzzy rules will be used in the inference engine

For this example, let us say that the input and output variables are divided into the following fuzzy sets:

**Table 6.6**: *Fuzzy Sets for the Example*

| Name | Type | Fuzzy sets |
|---|---|---|
| Temperature | Input | Cold, Normal, Hot |
| Humidity | Input | Low, Normal, High |
| Control | Output | Small, Medium, Large |

Let us use triangular membership functions for all the variables as shown in Figure 6.17.

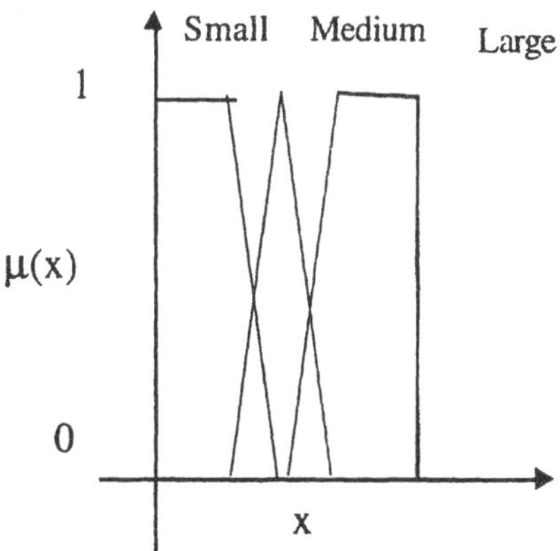


**Figure 6.17**: *Example Membership Functions*

Let us also assume the following two rules in the inference engine:
(7) If Temperature is Cold AND Humidity is High Then Control is Small
(8) If Temperature is Normal AND Humidity is Normal Then Control is Medium

Operation:
The operation follows the steps of Figure 6.16. These are described below:

Fuzzification: Let us assume that at some sampling time the measured temperature is x and the measured humidity is y. Then, we find out the value of the membership functions of the input fuzzy sets. As shown in Figure 6.18, let us say that the value for Cold (Temperature) is 0.48, the value for High (Humidity) is 0.76, that the value for Normal (Temperature) is 0.51, and the value for Normal (Humidity) is 0.51. We will use *singleton fuzzification*, using which we define new fuzzy set "Cold Temperature" as a fuzzy set with the following membership function

$$\mu_{\text{Cold Temperature}}(X) = \begin{cases} 0.76 & \text{if } X = x \\ 0 & \text{if } X \neq x \end{cases} \qquad (62)$$

Similarly for "High Humidity", we define:

$$\mu_{\text{High Humidity}}(X) = \begin{cases} 0.48 & \text{if } Y = y \\ 0 & \text{if } Y \neq y \end{cases} \qquad (63)$$

We can define fuzzy sets similarly for Normal Temperature and Normal Humidity.

Fuzzy Processing: Now, we use min function for "and" and also for implication and we use (56) for composition, the fuzzy output firing the first rule is shown in the first row of Figure 6.18. Similarly, the second rule gives us the second row. We combine the output of the two rules using "or" (union) operation and we have used the max operation for the t-norm here. The result is shown in the third row.

Defuzzification: Now the system needs to provide a crisp output control. This can be obtained by using different methods of defuzzification. In this example we choose it to be the x-axis point of the center of gravity of the fuzzy membership curve as shown in Figure 6.18.

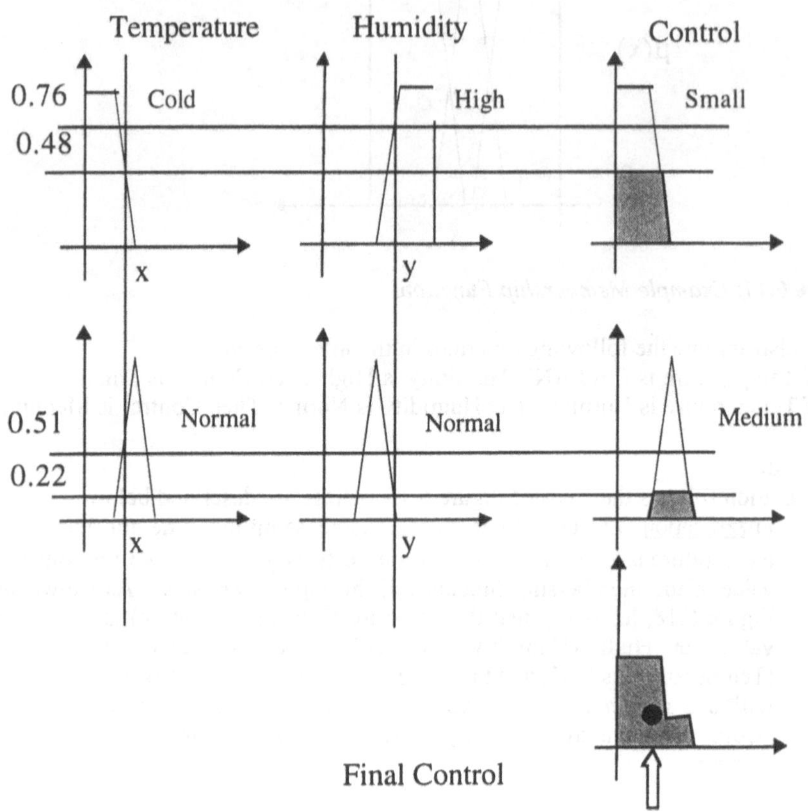


**Figure 6.18**: *Example Fuzzy Calculations*

In summary, we have the following:

The building block of the Fuzzy Logic Inference Engine (FLIE) is Modus Ponens, which states:

| | |
|---|---|
| Premise1: | x is A |
| Premise 2: | IF x is A THEN y is B |
| Consequence: | y is B |

This can also be expressed as: $(p \cap (p \rightarrow q)) \rightarrow q$. In fuzzy logic, this can be constructed by:

$$\mu_{B*}(y) = \sup_{x \in A*}[\mu_{A*}(x) * \mu_{A \to B}(x, y)] \tag{64}$$

In this sup-norm composition, we can use product or minimum as the t-norm for the star composition. Note that we are using the symbol * with sets A and B, this is because, as shown in the example, we construct new fuzzy set from set A and also obtain a new fuzzy set from B and both the new fuzzy sets have different membership functions. Now, if we use a singleton fuzzifier, then (64) becomes

$$\mu_{B*}(y) = \mu_{A*}(x') * \mu_{A \to B}(x, y) \tag{65}$$

where $\mu(x') = 1$, at x=x', and zero everywhere else, denoting the use of singleton fuzzification process. Since, the support of A* is x', we obtain

$$\mu_{B*}(y) = \mu_{A \to B}(x', y) \tag{66}$$

When there are multiple antecedents in a rule, such as, IF $u_1$ is $F_1^\ell$ and $u_2$ is $F_2^\ell$...and $u_m$ is $F_m^\ell$, THEN v is $G^\ell$, we can write

$$\mu_{A \to B}(x, y) = \mu_{F_1^\ell}(x_1) * \mu_{F_2^\ell}(x_2) * \ldots * \mu_{F_m^\ell}(x_m) \tag{67}$$

In order to connect the fuzzy rules, t-conorm, $\oplus$ will be used. An effective t-conorm is the max operation. Hence, the fuzzy output of the FLIE, in which there are p rules, is

$$\mu_{R1}(x, y) \oplus \mu_{R2}(x, y) \oplus \ldots \oplus \mu_{Rp}(x, y) \tag{68}$$

We use the singleton fuzzifier method for the fuzzification. A fuzzy singleton has support $x_a$, i.e., $\mu_{A*}(x_A) = 1$, for $x = x_A$ and $\mu_{A*}(x_A) = 0, \forall\, x \in U, (x \neq x_A)$.

There are many ways of performing defuzzification, such as

(1) Mean of Maximum Defuzzifier: In this method, the output of the defuzzifier acting on a variable x of a fuzzy set A is the mean of those crisp values of x, which give the maximum value of $\mu_A(x)$.

(2) Centroidal Method: In this method, the output of the defuzzifier acting on a variable x of a fuzzy set A is that crisp value of x, at which the centroid of the area under the membership function curve exists.

# 3. Sample Problem

In order to illustrate the ideas discussed above, we have designed a feedback control system for a simple network consisting of three alternate routes. Similar feedback controllers can be designed for larger and more complex traffic networks. For simplicity, we are assuming that each of the alternate route is just a single discrete section.

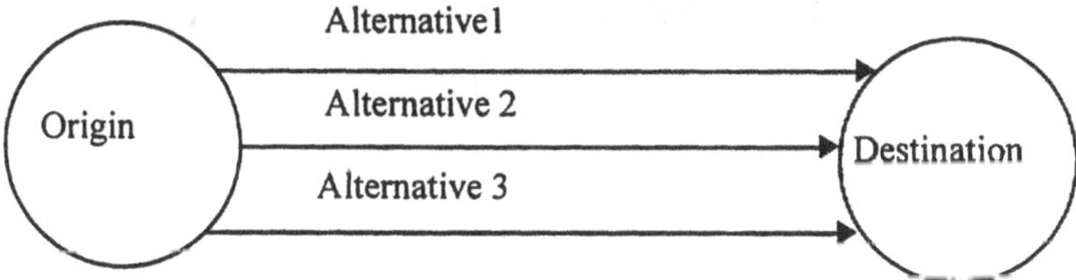


**Figure 6.19** *Test Network*

Although the traffic network presented in this chapter seems to be simplistic, it represents one of the most realistic settings for real-time traffic routing. The main goal of traffic routing for today's applications is to alleviate non-recurrent traffic congestion on a freeway by diverting freeway traffic to alternate routes. For most of the urban traffic networks, number of feasible alternate routes is not more than three. Moreover, due to the difficulties related to the real-time control of traffic at multiple locations, point diversion appears to be a practical near-term alternative to the network wide routing where more advanced traffic control as well as more complex dynamic routing algorithms might be needed. Thus, the sample problem presented in this chapter describes a very realistic scenario for demonstrating the feasibility and advantages of fuzzy feedback control for dynamic traffic routing. On the other hand, it is quite clear that the design of more advanced controllers that can tackle more complex networks and scenarios will be needed for future applications.

An incident that occurs on a freeway can be used as a perfect scenario to illustrate the realism of the sample problem. Now, let us assume that as a result of the incident, part of freeway traffic will be diverted to the alternate highways. Diversion will then be initiated at a point before the incident location and the traffic will be diverted back to the major freeway at a point past the incident. Every day occurrences of incident situations similar to the above example are abundant in Northern Virginia and other parts of the country. Given the real-time traffic control capabilities of most of the DOTs, a system that will regulate the point diversion using a widely used actuation method such as variable message signs appears to be a viable solution. However, in the future, with the widespread use of in-vehicle guidance devices the need for more complex controllers will be inevitable.

**System Dynamics**

The flow equations used for the three alternate routes are:

$$\rho_1(k+1)=\rho_1(k)-\frac{T}{\delta_1}[q_1^{out}(k)-q_1^{in}(k)] \qquad (69)$$

$$\rho_2(k+1)=\rho_2(k)-\frac{T}{\delta_2}[q_2^{out}(k)-q_2^{in}(k)]$$

$$\rho_3(k+1)=\rho_3(k)-\frac{T}{\delta_3}[q_3^{out}(k)-q_3^{in}(k)]$$

where

$$q_1^{out}(k)=\rho_1(k)v_1(k) \qquad (70)$$

$$q_2^{out}(k)=\rho_2(k)v_2(k)$$

$$q_3^{out}(k)=\rho_3(k)v_3(k)$$

$$v_1 = v_{f1}(1 - \frac{\rho_1}{\rho_{max1}}) \tag{71}$$

$$v_2 = v_{f2}(1 - \frac{\rho_2}{\rho_{max2}})$$

$$v_3 = v_{f3}(1 - \frac{\rho_3}{\rho_{max3}})$$

and

$$q_1^{in}(k) = \beta_1(k)U(k) \qquad 0 \le \beta_1(k) \le 1 \tag{72}$$

$$q_2^{in}(k) = \beta_2(k)U(k) \qquad 0 \le \beta_2(k) \le 1$$

$$q_3^{in}(k) = [1 - \beta_1(k) - \beta_2(k)]U(k)$$

These variables have been defined in Chapter 3. We have considered a very simple first order travel time function, which is obtained by dividing the length of a section by average velocity of vehicles on it. According to that, we have

$$\chi_1(k) = d_1 / [v_{f1}(1 - \frac{\rho_1}{\rho_{max1}})]$$

$$\chi_2(k) = d_2 / [v_{f2}(1 - \frac{\rho_2}{\rho_{max2}})]$$

$$\chi_3(k) = d_3 / [v_{f3}(1 - \frac{\rho_3}{\rho_{max3}})] \tag{73}$$

The state variables are the $\rho$'s for each route. This is considered a full state measurement problem, assuming the flows are measured, which can be converted into state variable values using the deterministic relationships. The control variables are the splitting rates. For this example, full compliance of the traffic flow to these splitting rates is assumed. This assumes that some technique of making the vehicles follow the splitting rate, such as Variable Message Signs (VMS) or in-vehicle communication is employed. The overall system in a standard nonlinear state-space form can be written as

$$\mathbf{x}(k+1) = \mathbf{f}(\mathbf{x}(k)) + \mathbf{g}(\mathbf{x}(k))\mathbf{u}(k) \tag{74}$$

$$\mathbf{y}(k) = \mathbf{h}(\mathbf{x}(k))$$

where

$$\mathbf{x}(k) = [\rho_1(k), \rho_2(k), \rho_3(k)]^T,$$

$$\mathbf{y}(k) = [\chi_1(k), \chi_2(k), \chi_3(k)]^T, \ \mathbf{u}(k) = [\beta_1, \beta_2]^{T},$$

$$\mathbf{f}(\mathbf{x}(k)) = \begin{bmatrix} \rho_1(k) - T\rho_1(k)v_{f1}(1 - \rho_1(k)/\rho_{max1})/\delta_1 \\ \rho_2(k) - T\rho_2(k)v_{f2}(1 - \rho_2(k)/\rho_{max2})/\delta_2 \\ \rho_3(k) - T\rho_3(k)v_{f3}(1 - \rho_3(k)/\rho_{max3})/\delta_3 + U(k) \end{bmatrix},$$

$$\mathbf{g}(\mathbf{x}(k)) = \begin{bmatrix} TU(k)/\rho_1(k) & 0 \\ 0 & TU(k)/\rho_2(k) \\ -TU(k)/\rho_1(k) & -TU(k)/\rho_2(k) \end{bmatrix}, \tag{75}$$

$$\mathbf{h}(\mathbf{x}(k)) = \begin{bmatrix} d_1/[v_{f1}(1-\frac{\rho_1(k)}{\rho_{max1}})] \\ d_2/[v_{f2}(1-\frac{\rho_2(k)}{\rho_{max2}})] \\ d_3/[v_{f3}(1-\frac{\rho_3(k)}{\rho_{max3}})] \end{bmatrix}$$

Here the system is in the local coordinates for the smooth state space manifold M, **f** is the smooth drift vector field on M, **g** is the smooth input vector field on M, and **h** is the smooth output vector field on M.

**Simple Fuzzy Feedback Control Law**

Although the dynamics of the system are described in the previous section, it is difficult to design analytic controllers which are robust to perturbations to a nonlinear plant which represent the real physical system. Fuzzy control provides a design methodology, which might prove effective in design of robust controllers for such systems. The details of the fuzzy controller used for this example are shown below.

The error terms to drive the controller are defined as

$$e_1(k) = \chi_3(k) - \chi_1(k) \tag{76}$$
$$e_2(k) = \chi_3(k) - \chi_2(k)$$

The integral term in the fuzzy controller is taken as error summation, so that

$$ie_1(k) = \sum_{p=1}^{k} \{\chi_3(p) - \chi_1(p)\}$$
$$ie_2(k) = \sum_{p=1}^{k} \{\chi_3(p) - \chi_2(p)\} \tag{77}$$

where p is the dummy time variable used for summation for discrete integration.

The fuzzy control we designed emulates the concept of PI controller. Structure of a PI controller with constant feedforward is shown below which would try to equilibrate travel times on alternate routes. We use fuzzy control to perform all the functions of feedback, feedforward, and the saturation. This controller is presented here to show the feasibility of fuzzy feedback control to meet the objective of achieving equal travel time in all the alternate routes, and in general to demonstrate the applicability of fuzzy feedback control to DTR and in general to DTA.

$$\beta_1(k) = \max[0, \min\{1, (1/3 + k_1 e_1 + k_{i1} ie_1)\}] \quad (78)$$

$$\beta_2(k) = \max[0, \min\{1 - \beta_1(k), (1/3 + k_2 e_2 + k_{i2} ie_2)\}]$$

where $k_1, k_{i1}, k_2,$ and $k_{i2}$ are the controller gains.

There are seven membership functions each for $e_1, e_2, ie_1,$ and $ie_2$, namely, negative high, negative medium, negative low, zero, positive low, positive medium, and positive high. These are shown in Figure 6.20.

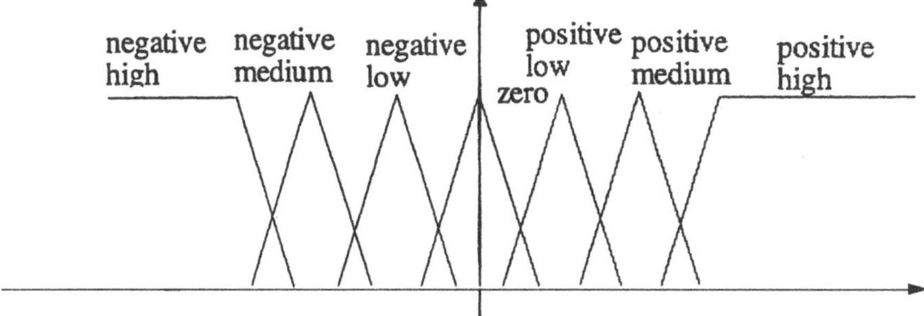


**Figure 6.20.** *Error Membership functions*

The support of each triangular membership function is divided into three equal parts. The triangles intersect one third at the bases of each others as shown in Figure 6.20. By putting this constraint, and the constraint that all the membership functions have the same magnitude of angles with the base, leave two degrees of freedom to divide the number line into membership functions. The two degrees of freedom are the magnitude of the base angle of the membership functions '$\theta$', and the number of membership functions 'n'. For the simulation purposes, we have $\theta$ = 1.24 radians and n=7.

For $\beta_1$ and $\beta_2$, we have three membership functions each, low, medium and high, as shown in Figure 6.21. Here, we use isosceles triangular membership functions covering the entire universe of discourse, which is the closed set [0, 1].

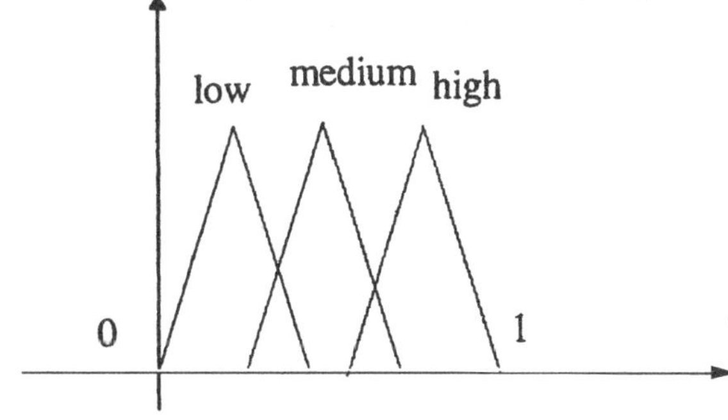


**Figure 6.21.** *Membership functions for* $\beta_1$ *and* $\beta_2$.

For this example, we have used 98 rules utilizing all the possible combinations of fuzzy values of the errors. The rules dealing with $e_1$ and $ie_1$ are 49 (7 X 7), and similarly rules dealing with $e_2$ and $ie_2$ are 49 too, making the total 98. Note that this is a large set of rules for this choice. The reasons for that are, we have not used a default value; there is region in every membership function, which has no overlap; and we have used the min function in combining the membership functions for fuzzy processing. The number of rules can be drastically reduced by changing these options. The following are a few representative rules utilized.

IF $e_1$ is positive low, AND $ie_1$ is positive low, THEN $\beta_1$ is low.
IF $e_2$ is positive low, AND $ie_2$ is positive low, THEN $\beta_2$ is low.
IF $e_1$ is negative high, AND $ie_1$ is negative low, THEN $\beta_1$ is low.
IF $e_2$ is negative high, AND $ie_2$ is negative low, THEN $\beta_2$ is low.
IF $e_1$ is positive high, AND $ie_1$ is negative low, THEN $\beta_1$ is medium.
IF $e_2$ is positive high, AND $ie_2$ is negative low, THEN $\beta_2$ is medium.
IF $e_1$ is positive high, AND $ie_1$ is negative medium, THEN $\beta_1$ is low.
IF $e_2$ is positive high, AND $ie_2$ is negative medium, THEN $\beta_2$ is low.
IF $e_1$ is positive medium, AND $ie_1$ is positive medium, THEN $\beta_1$ is medium.
IF $e_2$ is positive medium, AND $ie_2$ is positive medium, THEN $\beta_2$ is medium.

In this example, we use singleton fuzzification, and centroid defuzzification for crisp-fuzzy and fuzzy-crisp conversions.

**Results & Description for different scenarios**

For the simple example problem and its feedback control solution, we have performed several test runs for different scenarios. We have tested the simulation using three scenarios. For each scenario, we have considered a simple network that consists of three alternate routes. The splitting decisions are made at one decision point only. Alternate routes have different free flow travel times. The one with the lowest travel time can be assumed to be the freeway and two others with higher travel times can be considered to be highways with lower level of service. Brief description of each scenario and the corresponding plots are shown below.

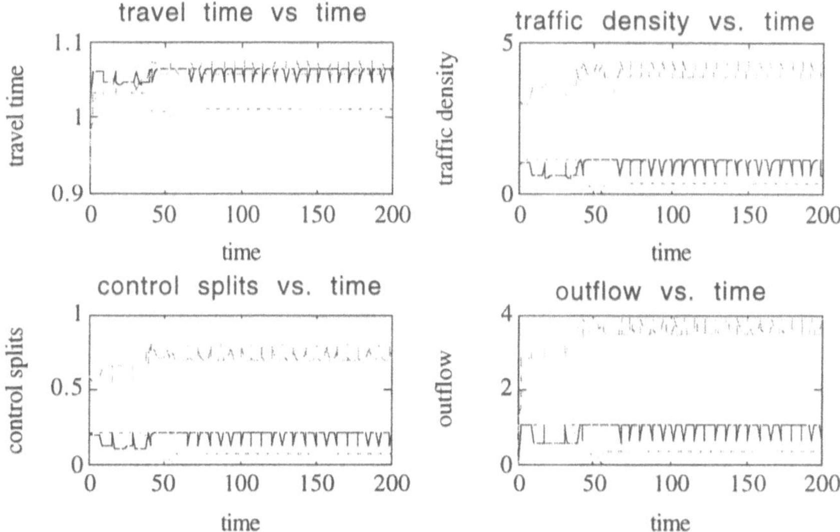


**Figure 6.22**. *Scenario 1*

*First Scenario:*

In this scenario, we have constant inflow and no congestion for all the alternate routes. This scenario represents normal traffic conditions (Figure 6.22). As it can be seen in the first plot, travel times become equal with some steady state error and some steady state oscillations. These errors can be further reduced by re-tuning the fuzzy control.

*Second Scenario:*

This scenario simulates congestion, on route 2 from time 50 to 75, and relief on route 3, from time 125 to 150 (Figure 6.23). The congestion on route 2 may be due to a temporal bottleneck caused by an incident and the relief on route 3 may be due to the clearance of an incident that existed before. This fuzzy control is not very responsive to the change in congestion level. It can also be improved by further tuning of the control parameters.

*Scenario 3:*

This last scenario has a sinusoidal inflow (demand) function and same traffic patterns as scenario 2 (Figure 6.24). The fluctuations in inflow traffic modeled by the sinusoidal are meant to represent the natural traffic fluctuations. Here, the steady state errors are reduced, but there is high control activity, which can be filtered out.

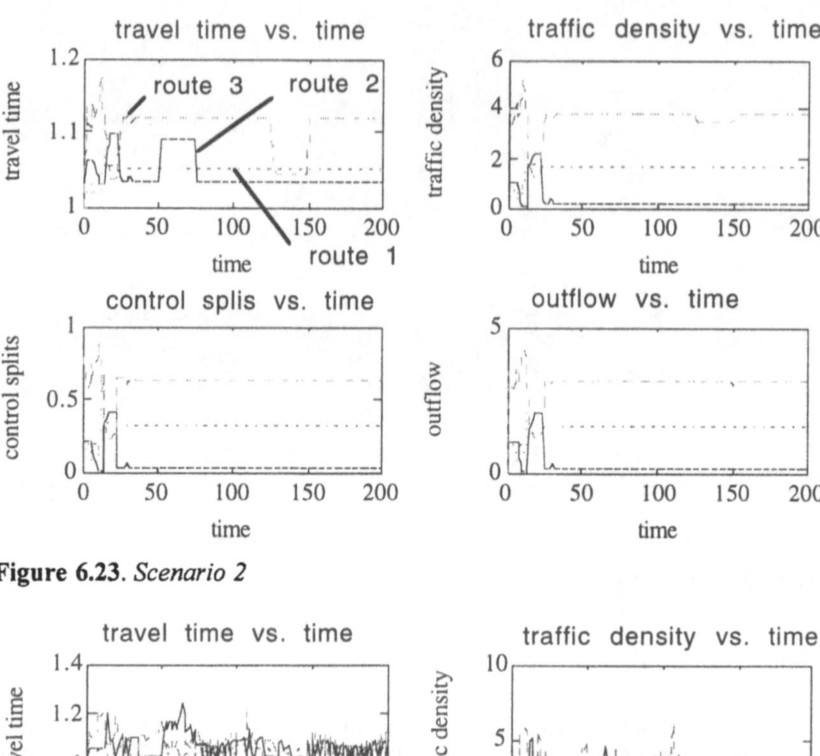


**Figure 6.23.** *Scenario 2*

**Figure 6.24.** *Scenario 3*

# 4. Summary

- A review of fuzzy logic methods was provided and a simple but illustrative example shown.
- In the simulation example, we have shown that fuzzy feedback control is viable and attractive solution to the on-line dynamic traffic control/routing problem. In order to run the feedback control, we obtained desired states that

the controller will track from the traffic sensors. In this control problem, the travel time to be tracked for each route was the travel time of the next route. This solution worked nicely for this simple problem under different demand and traffic conditions. We can try to generalize this concept for more complex networks and conduct more research to design tracking traffic controller.

# 5. Exercises

## Questions

Question 1:What is the relationship between fuzzy logic and probability theory?
Question 2:Compare briefly set theory, propositional logic and Boolean algebra.
Question 3: Which set is a subset of every set?
Question 4: Which set contains every set as its subset?
Question 5: What is the range of membership functions of crisp sets?
Question 6: What is the range of membership functions of fuzzy sets?
Question 7: How is the set R(A,B) related to the product space A×B?
Question 8: State Modus Ponens
Question 9: How is partial ordering defined in Boolean algebra?
Question 10: What does the following mean for a fuzzy set F

F=0.1/2+0.45/4+0.12/6+0.2/8

Plot the membership function for this set.
Question 11: What are the most common membership functions for fuzzy unions, fuzzy intersections, and fuzzy complements?
Question 12: What is a sup-star composition on fuzzy relations? If the first relation is a fuzzy set in a sup-star composition, how does it effect the result?
Question 13: What is singleton fuzzification?
Question 14: How is defuzzification performed in fuzzy systems?
Question 15: Show how a fuzzy control can be designed for a network level traffic routing problem.

## Problems

*Problem 1*

Using Venn diagrams for sets A and B, prove that $\overline{A \cup B} = \overline{A} \cap \overline{B}$

*Problem2*

Write down R(A,B), R(B,C) and R(A,C) in matrix form for the problem shown in the following figure.

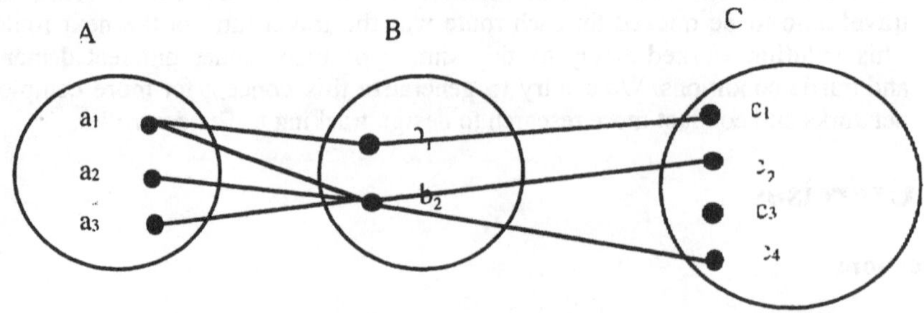


Show that relations (27) and (31) are valid for this example.

*Problem3*

Using a truth table prove a•(b+c)=(a•b)+(a•c)

*Problem4*

Perform the union and intersection of fuzzy sets (39) and (40) and draw their membership functions. Also perform the complements of A and B and show their membership functions.

*Problem5*

Use p as a singleton fuzzy set in (59) and (60) and then calculate and draw the membership function of the implication in (59) and (60) using min for intersection, max for union, and (50) for complement.

*Problem6*

Redo the example of Figure 6.18 by using the multiplication operator instead of min operation for "AND" as well as for implication.

## 6. References


(1) Zadeh, L. A., Fuzzy Sets, *Information Control*, 8(3)(1965)338-353.
(2) Zadeh, L. A., The Role of Fuzzy Logic in the Management of Uncertainty in Expert Systems, *Fuzzy Sets and Systems*, 11(1985) 199-227.
(3) Zadeh, L. A., The Concept of Linguistic Variable and its Application to Approximate Reasoning - I, II, III, *Information Sciences*, 8(1975) 199-249, 8(1975) 301-357, 9(1975) 43-80.
(4) Mamdani, E. H., Advances in the Linguistic Synthesis of Fuzzy Controls, *Int. J. Man-Machine Studies*, 8(1976) 699-678.
(5) Mamdani, E. H., J. J. Ostergaard, and E. Lembessis, Use of Fuzzy Logic for Implementing Rule-Based Control of Industrial Processes, Wang, P. P. and Chang, S. K., eds. *Advances in Fuzzy Sets Possibility Theory and Application*, New York: Plenum(1983).

(6) Kickert, W. J. M., and H. R.Van Nauta Lemke, Application of a Fuzzy Controller in a Warm Water Plant, *Automatica*, 12(1976) 301-308.

(7) Jain, P., and A. Rege, Survey of U.S. Applications of Fuzzy Logic:Hardware, Controls, Expert Systems, Patent Recognition, and Others, Agogino Engineering, 130 Wilding Ln., Oakland, CA, 94618(1987).

(8) Lotan T. and H. N. Koutsopoulos, "Fuzzy Control and Approximate Reasoning Models for Route Choice in the Presence of Information", Paper presented at the 73rd Annual TRB Meeting, 1993.

(9) Lotan T. and H. N. Koutsopoulos, "Approximate Reasoning Models for Route Choice in the Presence of Information", Transportation and Traffic Theory, Proceedings of the 12th International Symposium on the Theory of Traffic Flow and Transportation, Berkeley, California, 21-23 July 1993.

(10) Sasaki, T, and T. Akiyama, ' Traffic Control Process of Expressway by Fuzzy Logic", Fuzzy Sets and Systems 26 165-178, 1988.

(11) Chen, L. C., A. D. May, and D. M. Auslander, " Freeway Ramp Control Using Fuzzy Set Theory for Inexact Reasoning", Transpn. Res.-A, Vol. 24 A, No. 1., p. 15-25, 1990.

(12) Mendel, Jerry M., 'Fuzzy logic systems for engineering: a tutorial', Proc. of the IEEE, vol. 83, No. 3, March 1995.

(13) Klir, George J. and Folger, Tina A., *Fuzzy Sets, Uncertainty, and Information*, Prentice Hall, 1988.

# CHAPTER 7
# FEEDBACK CONTROL FOR DYNAMIC TRAFFIC ROUTING IN LUMPED PARAMETER SETTING

## Objectives

- To solve the point diversion problem for n routes between two nodes using feedback control
- Perform simulations to show the results of the controller

## 1. Introduction

In this chapter, dynamic traffic routing problem is formulated as a feedback control problem that determines the time-dependent split parameters at the diversion point for routing the incoming traffic flow onto the alternate routes in order to achieve a user-equilibrium traffic pattern. Feedback linearization and sliding mode control techniques are used to solve this specific user-equilibrium formulation of the dynamic traffic routing problem. The control input is the traffic split factor at the diversion point. By transforming the dynamics of the system into canonical form, a control law is obtained which cancels the nonlinearities of the system in the case of feedback linearization, and in the case of sliding mode control there is an additional term for cancellation of partially known uncertainties. Simulation results show that the performance of these controllers on a test network is quite promising.

**Notation**

*Traffic Variables*

| | |
|---|---|
| $q_i$ | traffic volume entering link i |
| $q_{i,j}$ | traffic volume entering link j of route i |
| $\rho_i$ | traffic density in link i |
| $\rho_{i,j}$ | traffic density in link j of route i |
| $v_i$ | average traffic speed in link i |
| $r_i$ | ramp traffic flow entering link i |
| $s_i$ | ramp traffic flow exiting link i |
| $\beta$ | traffic split factor at a node |
| $U$ | input flow at a node |
| $\mathbf{x}$ | state vector |
| $\mathbf{y}$ | measurement vector |
| $\mathbf{u}$ | input vector |

| | |
|---|---|
| **e** | error vector |
| J | objective function |
| $\chi(.,.)$ | travel time function |

*Section Parameters*

| | |
|---|---|
| $\rho_{max\,i}$ | traffic jam density for link i |
| $v_{fi}$ | freeflow traffic speed for link i |
| $k_i$ | travel time parameter for link i |
| $k_{ijp}$ | travel time parameter for link j route i |
| $v_{fij}$ | freeflow traffic speed for link j route i |

*Others*

| | |
|---|---|
| t | time |
| $t_f$ | final time for finite horizon optimal control problems |
| $L_f h$ | Lie derivative of scalar h with respect to vector field **f** |
| $y_i^{\gamma_j}$ | derivative of the order $\gamma_i$ of $y_i$ |
| $\eta_i$ | state variable for internal dynamics |
| **A** | transition matrix |
| **B** | decoupling matrix |
| F | transition scalar |
| G | decoupling scalar |

## 2. System Dynamics and DTR Problem

For modeling, the highway is subdivided into several sections, as shown in Figure 7.1.

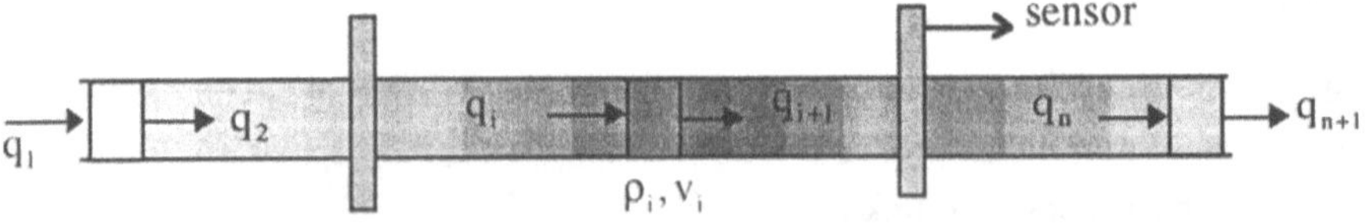


**Figure 7.1.** *Highway divided into sections.*

The following Ordinary Differential Equations (ODE) with the given variable relationships can be used to model a space discretized freeway:

$$\frac{d}{dt}\rho_i = \frac{1}{\delta_i}[q_i(t) - q_{i+1}(t) + r_i(t) - s_i(t)], \quad i = 1,2,...n. \tag{1}$$

$$q_i(t) = \rho_i(t)v_i(t) \tag{2}$$

$$v_i = v_{fi}(1 - \frac{\rho_i}{\rho_{max\,i}}) \tag{3}$$

Here, $r_i(t)$ and $s_i(t)$ terms indicate the on-ramp and off-ramp flows, $\rho_i$ is the density of the traffic as a function of x and time t, $q(t)$ is the flow at given x, (t) $v_{fi}$ is the free flow speed, and $\rho_{max\,i}$ is the jam density. Equation (1) and the output equations (4) give the mathematical model for a highway, which can be represented in a standard nonlinear state space form for control design purposes.

$$y_j = g_j(\rho_1, \rho_2, \ldots, \rho_n), \quad j = 1, 2, \ldots, p, \tag{4}$$

The standard state space form is

$$\begin{aligned} \frac{d}{dt}\mathbf{x}(t) &= \mathbf{f}[\mathbf{x}(t), \mathbf{u}(t)], \\ \mathbf{y}(t) &= \mathbf{g}[\mathbf{x}(t), \mathbf{u}(t)], \\ \mathbf{x}(0) &= \mathbf{x}_0, \end{aligned} \tag{5}$$

where $\mathbf{x} = [\rho_1, \rho_2, \ldots, \rho_n]^T$ and $\mathbf{u}(t) = q_0(t)$.

The DTR problem for n-alternate routes is described in section 5.1 of this book. Please revise that section before reading the next section.

In the discretized traffic flow model, the freeway is divided into sections with aggregate traffic densities [1]. Sensors are used to measure variables such as densities, traffic flow, and traffic average speeds in these sections, which can be used by the feedback controller to give appropriate commands to actuators like VMS and HAR.

In the next section, we present a user-equilibrium formulation of the DTR problem for the discretized traffic flow model given in section 2. Then, in the following sections, three different controllers for three different versions of the same DTR problem formulation are developed using the feedback linearization technique.

**DTR Formulation**

We first present a DTR formulation for the two alternate routes problem which is then generalized for n routes case. The two routes are divided into $n_1$ and $n_2$ sections, respectively. For simplicity, we are considering the static velocity relationship, and ignoring the effect of downstream flow. Hence, the model used is

$$\frac{d}{dt}\rho_{i,j} = \frac{1}{\delta_i}[q_{i,j-1}(t) - q_{i,j}(t)], \tag{6}$$

$$(i,j) = ((1,1),(1,2),\ldots,(1,n_1),(2,1),(2,2),\ldots,(2,n_2))$$

with relationships (2) and (3). The control input is given by

$$\begin{aligned} \beta(t)U(t) &= q_{1,0}(t), \quad 0 \le \beta \le 1, \\ (1-\beta(t))U(t) &= q_{2,0}(t) \end{aligned} \tag{7}$$

The flow U(t) is measured as a function of time, and the splitting rate $\beta(t)$ is the control input. The output measurement could be the full state vector, i.e., vector of flows of all the sections, or a subset of that. The control problem can be stated as: find $\beta_o(t)$, the optimal $\beta(t)$, which minimizes

$$J(\beta)=\int_0^{t_f}[\sum_{i=1}^{m}\chi(\rho_i)-\sum_{m+1}^{m+p}\chi(\rho_j)]^2 dt \tag{8}$$

where $\chi(.,.)$ is the travel time function and $t_f$ is the final time. Note that a feedback solution is needed for the problem, not an open loop optimal control. Hence, we can either decide the structure of the feedback control, such as a PID control with constant gains, and solve numerically for the optimal values of the gains, or we can state the control objective for a standard feedback control problem, such as steady state asymptotic stability given by

$$Lt_{t\to\infty}[\sum_{i=1}^{m}\chi(\rho_i)-\sum_{m+1}^{m+p}\chi(\rho_j)]\to 0 \tag{9}$$

and some transient behavior characteristics such as a specified settling time or percent overshoot.

Now we formulate the same DTR formulation for a generalized case, for an n alternate route problem, as follows:

<u>Problem</u>: Find $\beta_o^i$, i=1,2,...,n, which minimize

$$J(\beta_0^i, i=1,2,..n)=\int_0^{t_f}[\sum_{k,p}\{\sum_{i=1}^{n_k}\chi(\rho_{i,k})-\sum_{j=1}^{n_p}\chi(\rho_{j,p})\}^2 dt \tag{10}$$

(k=1,2,...,n, p=1,2,...,n, and the summations are taken over the total number of combinations of n and p, and not permutations, so that (k,p)=(1,2) is considered the same as (k,p)=(2,1), and hence only one of these two will be in the summation), or which guarantee

$$Lt_{t\to\infty}\mathbf{e}\to\mathbf{0}, \quad \text{where} \tag{11}$$

$$\mathbf{e}=[\{\sum_{i=1}^{n_1}\chi(\rho_{i,1})-\sum_{j=1}^{n_2}\chi(\rho_{j,2})\},...,\{\sum_{i=1}^{n_k}\chi(\rho_{i,k})-\sum_{j=1}^{n_p}\chi(\rho_{j,p})\},...]$$

with some transient behavior characteristics like a specified settling time or percent overshoot for the system

$$\frac{d}{dt}\rho_{i,j}=\frac{1}{\delta_i}[q_{i,j-1}(t)-q_{i,j}(t)], \quad (i,j)=((1,1),...,(1,n_1),(2,1),...,(2, \tag{12}$$

with given full and partial state observation, and input constraints

$$\sum_{i=1}^{n}q_{i,0}(t)=U(t) \text{ and } \sum_{i=1}^{n}\beta^i=1. \tag{13}$$

# 3. Feedback Linearization Technique

Feedback linearization is an appropriate technique for developing feedback controllers for nonlinear systems similar to the DTR model described above. The feedback linearization technique is applicable to an input affine square multiple input multiple output (MIMO), system. The details on exact nonlinear decoupling technique (feedback linearization) can be found in [2, 3, 4], and are briefly summarized here for the DTR application. Let us consider the following square MIMO system:

$$\dot{\mathbf{x}}(t) = \mathbf{f}(\mathbf{x}) + \sum_{i=1}^{p} g_i(\mathbf{x})u_i \qquad (14)$$
$$y_j = h_j(\mathbf{x}) \quad j = 1,2,\ldots,p$$

This can be written in a compact form as

$$\Sigma: \begin{array}{c} \dot{\mathbf{x}}(t) = \mathbf{f}(\mathbf{x}) + \mathbf{g}(\mathbf{x})\mathbf{u} \\ \mathbf{y} = \mathbf{h}(\mathbf{x}) \end{array} \qquad (15)$$

where, $\mathbf{x} \in R^n, \mathbf{f}(\mathbf{x}): R^n \to R^n$, $\mathbf{g}(\mathbf{x}): R^p \to R^n$, $\mathbf{u} \in R^p$, and $\mathbf{y} \in R^p$. The vector fields of $\mathbf{f}(\mathbf{x})$ and $\mathbf{g}(\mathbf{x})$ are analytic functions.

It is assumed that for the system $\Sigma$, each output $y_j$ has a defined relative degree $\gamma_j$. The concept of relative degree implies that if the output is differentiated with respect to time $\gamma_j$ times, then the control input appears in the equation. This can be succinctly represented using Lie derivatives. A definition of a Lie derivative is given below, after which the definition of relative degree in terms of Lie derivatives is stated.

*Definition* (Lie Derivative): Lie derivative of a smooth scalar function $h: R^n \to R$ with respect to a smooth vector field $\mathbf{f}: R^n \to R^n$ is given by $L_f h = \frac{\partial h}{\partial \mathbf{x}}\mathbf{f}$. Here, $L_f h$ denotes the Lie derivative of order zero. Higher-order Lie derivatives are given by $L_f^i h = L_f(L_f^{i-1}h)$.

*Definition* (Relative Degree): The output $y_j$ of the system $\Sigma$ has a relative degree $\gamma_j$ if, $\exists$ an integer, s.t. $L_{g_i} L_f^{\ell} h(\mathbf{x}) \equiv 0 \quad \forall \ell < \gamma_j - 1, \forall 1 \le i \le p, \forall x \in U$, and $L_{g_i} L_f^{\gamma_j - 1} h(\mathbf{x}) \ne 0$. $U \subset R^n$ which is in a given neighborhood of the equilibrium point of the system $\Sigma$. The total relative degree of the system r is defined to be the sum of the relative degrees of all the output variables, i.e., $r = \sum_{j=1}^{p} \gamma_j$.

By successively taking the Lie derivatives of each of the output variables up to their respective relative degrees, we obtain

$$\begin{bmatrix} y_1^{\gamma_1} \\ y_2^{\gamma_2} \\ \cdot \\ \cdot \\ \cdot \\ y_1^{\gamma_p} \end{bmatrix} = \begin{bmatrix} L_f^{\gamma_1} h_1(\mathbf{x}) \\ L_f^{\gamma_2} h_2(\mathbf{x}) \\ \cdot \\ \cdot \\ \cdot \\ L_f^{\gamma_p} h_p(\mathbf{x}) \end{bmatrix} + \begin{bmatrix} L_{g_1} L_f^{\gamma_1 - 1} h_1(\mathbf{x}) & \ldots & L_{g_p} L_f^{\gamma_1 - 1} h_1(\mathbf{x}) \\ L_{g_p} L_f^{\gamma_2 - 1} h_1(\mathbf{x}) & \ldots & L_{g_p} L_f^{\gamma_2 - 1} h_1(\mathbf{x}) \\ & \cdot & \\ & \cdot & \\ & \cdot & \\ L_{g_1} L_f^{\gamma_p - 1} h_1(\mathbf{x}) & \ldots & L_{g_p} L_f^{\gamma_p - 1} h_1(\mathbf{x}) \end{bmatrix} \mathbf{u} \tag{16}$$

This can be written as

$$\mathbf{y}^{\gamma} = \mathbf{A}(\mathbf{x}) + \mathbf{B}(\mathbf{x})\mathbf{u} \tag{17}$$

where

$$\mathbf{y}^{\gamma} = \begin{bmatrix} y_1^{\gamma_1} & y_2^{\gamma_2} & \ldots & y_1^{\gamma_p} \end{bmatrix}^T \tag{18}$$

$$\mathbf{A}(\mathbf{x}) = \begin{bmatrix} L_f^{\gamma_1} h_1(\mathbf{x}) & L_f^{\gamma_2} h_2(\mathbf{x}) & \ldots & L_f^{\gamma_p} h_p(\mathbf{x}) \end{bmatrix}^T \tag{19}$$

$$\mathbf{B}(\mathbf{x}) = \begin{bmatrix} L_{g_1} L_f^{\gamma_1 - 1} h_1(\mathbf{x}) & \ldots & L_{g_p} L_f^{\gamma_1 - 1} h_1(\mathbf{x}) \\ L_{g_p} L_f^{\gamma_2 - 1} h_1(\mathbf{x}) & \ldots & L_{g_p} L_f^{\gamma_2 - 1} h_1(\mathbf{x}) \\ & \cdot & \\ & \cdot & \\ & \cdot & \\ L_{g_1} L_f^{\gamma_p - 1} h_1(\mathbf{x}) & \ldots & L_{g_p} L_f^{\gamma_p - 1} h_1(\mathbf{x}) \end{bmatrix} \tag{20}$$

If the decoupling matrix $\mathbf{B}(\mathbf{x})$ is invertible, then we can use the feedback control law (21) to obtain the decoupled dynamics (22).

$$\mathbf{u} = (\mathbf{B}(\mathbf{x}))^{-1}[-\mathbf{A}(\mathbf{x}) + \mathbf{v}] \tag{21}$$

$$\mathbf{y}^{\gamma} = \mathbf{v} \tag{22}$$

where

$$\mathbf{v} = \begin{bmatrix} v_1 & v_2 & \ldots & v_p \end{bmatrix}^T$$

The vector **v** can be chosen to render the decoupled system (17) stable with desired transient behavior. Now, if the relative degree of the system r is less than the order of the system n, then the closed loop system should also have stable internal dynamics. In order to study that, one can define state variables $\eta_i(\mathbf{x}), i = 1, 2 \ldots, n - r$, which are independent of the state variables r related to the output of the system, and are also independent of each other. The internal dynamics of the system can then be written as:

$$\dot{\eta} = \mathbf{w}(\zeta, \eta) + \mathbf{P}(\zeta, \eta)\mathbf{u} \tag{23}$$

with (k=1,2,...,n-r) and (i=1,2,...,p)

$$w_k(\zeta, \eta) = L_f \eta_k(\mathbf{x}) \tag{24}$$

$$P_{ki}(\zeta,\eta) = L_{g_i}\eta_k(\mathbf{x}) \tag{25}$$

The feedback controller designed for (17) using the feedback linearization technique should also guarantee the stability of the internal dynamics described in (23). Note that for a single input case, we could use the fact that $L_g\eta_k(\mathbf{x}) = 0$ to choose the independent internal dynamics state variables, but for the multiple input case, this condition is not valid, unless the vectors of **g** are involutive. Note that the control law (21) has to satisfy the constraints (13). In case the constraints are not satisfied, the control variables take extreme values, and the desired performance of the eigenvalues is not achieved. The moment the traffic condition changes, such that the control variables belong to the feasible set, the performance of the system comes back to the desired state. As an example, to achieve a fast response time our controller might try to overcompensate, but in reality the constraints will produce a slower rate than desired by the design; nevertheless, the system will move towards equal travel times.

# 4. Sample Problem (Two Alternate Routes with One Section Each)

In order to illustrate the ideas discussed above, we have designed a feedback control law for the two alternate routes problem with a single section each. The control is based on feedback linearization technique for nonlinear systems. The technique is based on defining a diffeomorphism and performing the transformation on the state variables in order to convert them into the canonical form. A continuously differentiable mapping from space A to B which has a continuously differentiable inverse is called a diffeomorphism. If the relative degree of the system is less than the system order, then the internal dynamics are studied to ensure that it is stable. The details of this technique are given in [4]. In this problem, the system order is two and the relative degree is one.

The space discretized flow equations used for the two alternate routes are:

$$\dot{\rho}_1 = -\frac{1}{\delta_1}\left[ v_{f1}\rho_1(1-\frac{\rho_1}{\rho_{m1}}) - \beta U \right], \tag{26}$$

$$\dot{\rho}_2 = -\frac{1}{\delta_2}\left[ v_{f2}\rho_2(1-\frac{\rho_2}{\rho_{m2}}) + \beta U - U \right], \tag{27}$$

We have considered a simple first order travel time function, which is obtained by dividing the length of a section by the average velocity of the vehicles on it. According to that, the travel time can be calculated as

$$\chi_1(k) = d_1/[v_{f1}(1-\frac{\rho_1}{\rho_{m1}})], \tag{28}$$

$$\chi_2(k) = d_2/[v_{f2}(1-\frac{\rho_2}{\rho_{m2}})], \tag{29}$$

where $d_1$ and $d_2$ are section lengths, $v_{f1}$ and $v_{f2}$ are the free flow speeds of each section, and $\rho_{m1}$ and $\rho_{m2}$ are the maximum (jam) densities of each section. Since we need to equate the travel times according to the user-equilibrium DTR formulation discussed in the previous section, we take the new transformed state variable y as the difference in travel times. Differentiating the equation representing y in terms of the state variables introduces the input split factor into the dynamic equation. Therefore, that transformed equation can be used to design the input that cancels the nonlinearities of the system and introduces a design input $v$, which can be used to place the poles of the error equation for asymptotic stability. These steps are shown below.

The variable y is equal to the difference in the travel time on the two sections.

$$y = \frac{k_1}{(k_2 - \rho_1)} - \frac{k_3}{(k_4 - \rho_2)} \tag{30}$$

where $k_1 = \dfrac{d_1 \cdot \rho_{m1}}{v_{f1}}, k_2 = \rho_{m1}, k_3 = \dfrac{d_2 \cdot \rho_{m2}}{v_{f2}}, k_4 = \rho_{m2}$.

This equation can be differentiated with respect to time to give the travel time difference dynamics.

$$\dot{y} = \frac{k_1 \dot{\rho}_1}{(k_2 - \rho_1)^2} - \frac{k_3 \dot{\rho}_2}{(k_4 - \rho_2)^2} \tag{31}$$

By substituting (26) and (27) in (31), we obtain

$$\dot{y} = -\frac{k_1\left(v_{f1}\rho_1(1 - \frac{\rho_1}{\rho_{m1}}) - \beta U\right)}{\delta_1 (k_2 - \rho_1)^2} + \frac{k_3\left(v_{f2}\rho_2(1 - \frac{\rho_2}{\rho_{m2}}) + \beta U - U\right)}{\delta_2 (k_4 - \rho_2)^2} \tag{32}$$

This equation can be rewritten in the following form:

$$\dot{y} = F + G\beta \tag{33}$$

where

$$F = \left[ -\frac{k_1 v_{f1} \rho_1}{\delta_1 (k_2 - \rho_1)^2}(1 - \frac{\rho_1}{\rho_{m1}}) + \frac{k_3}{\delta_2 (k_4 - \rho_2)^2}\left((1 - \frac{\rho_2}{\rho_{m2}}) v_{f2} \rho_2 \right.\right. \tag{34}$$

$$G = \left( \frac{k_1}{\delta_1 (k_2 - \rho_1)^2} + \frac{k_3}{\delta_2 (k_4 - \rho_2)^2} \right) U \tag{35}$$

Hence, a feedback linearization control law can be designed to cancel the nonlinearities and provide the desired error dynamics. The feedback control law given in (21) is used,

$$\beta = G^{-1}(-F + v) \tag{36}$$

which gives the closed loop dynamics as

$$y = v \tag{37}$$

As was mentioned earlier, since the relative degree of the system is one, and the system order is two, we need to test the stability or boundedness of the second transformed state variable given by

$$\eta = \delta_1\rho_1 + \delta_2\rho_2 \tag{38}$$

The state variable $\eta$ is bounded since the densities on the sections cannot exceed the corresponding jam densities. This assumes that no traffic from ramps enters a section with jam density, and also that at the node, the traffic flow into the sections is zero if jam density is reached. This is a reasonable assumption since measured traffic density will never become higher than the jam density.

$$\eta \le \delta_1\rho_{m1} + \delta_2\rho_{m2} \tag{39}$$

and hence the overall system is exponentially stable $(y \to 0)$ if we choose $v = -Ky, \quad K > 0$, and y asymptotically goes to zero as $y(t) = y(0)e^{-Kt}$. This implies that when a splitting value based on (36) is utilized, the difference in travel time of two alternate routes will go to zero at an exponential rate. Hence, the closed loop traffic system controlled by the proposed feedback linearization law is exponentially stable and has desired transient behavior. Note that if input saturation occurs, the rate of convergence cannot be guaranteed.

# 5. Sample Problem (Two Alternate Routes with Two Sections Each)

Now, we extend the above problem to a case with two sections and follow the same steps for designing a new controller for this extended system. The space discretized flow equations used for the two alternate routes are:

$$\dot{\rho}_{11} = -\frac{1}{\delta_{11}}\left[ v_{f11}\rho_{11}(1 - \frac{\rho_{11}}{\rho_{m11}}) - \beta U \right], \tag{40}$$

$$\dot{\rho}_{12} = -\frac{1}{\delta_{12}}\left[ v_{f12}\rho_{12}(1 - \frac{\rho_{12}}{\rho_{m12}}) - v_{f11}\rho_{11}(1 - \frac{\rho_{11}}{\rho_{m11}}) \right], \tag{41}$$

$$\dot{\rho}_{21} = -\frac{1}{\delta_{21}}\left[ v_{f21}\rho_{21}(1 - \frac{\rho_{21}}{\rho_{m21}}) + \beta U - U \right] \tag{42}$$

$$\dot{\rho}_{22} = -\frac{1}{\delta_{12}}\left[ v_{f22}\rho_{22}(1 - \frac{\rho_{22}}{\rho_{m22}}) - v_{f21}\rho_{21}(1 - \frac{\rho_{21}}{\rho_{m21}}) \right] \tag{43}$$

We have considered a simple first order travel time function, which is obtained by dividing the length of a section by the average velocity of vehicles on it. According to that, we approximate travel time as

$$\chi_1(t)=\frac{d_{11}}{v_{f11}(1-\frac{\rho_{11}}{\rho_{m11}})}+\frac{d_{12}}{v_{f12}(1-\frac{\rho_{12}}{\rho_{m12}})}, \tag{44}$$

$$\chi_2(t)=\frac{d_{21}}{v_{f21}(1-\frac{\rho_{21}}{\rho_{m21}})}+\frac{d_{22}}{v_{f22}(1-\frac{\rho_{22}}{\rho_{m22}})}, \tag{45}$$

where $d_1$ and $d_2$ are section lengths, $v_{f1}$ and $v_{f2}$ are the free flow speeds of each section, and $\rho_{m1}$ and $\rho_{m2}$ are the maximum (jam) densities of each section.

The system can be written in the standard nonlinear input affine form

$$\dot{\mathbf{x}}(t)=\mathbf{f}(\mathbf{x},t)+\mathbf{g}(\mathbf{x},t)\mathbf{u}(t)$$
$$\mathbf{y}(t)=\mathbf{h}(\mathbf{x},t) \tag{46}$$

where

$$\mathbf{x}=[\rho_{11}\ \rho_{12}\ \rho_{21}\ \rho_{22}]',\mathbf{u}(t)=\beta \tag{47}$$

$$\mathbf{f}=\begin{bmatrix} -\frac{1}{\delta_{11}}\left[v_{f11}\rho_{11}(1-\frac{\rho_{11}}{\rho_{m11}})\right] \\ -\frac{1}{\delta_{12}}\left[v_{f12}\rho_{12}(1-\frac{\rho_{12}}{\rho_{m12}})-v_{f11}\rho_{11}(1-\frac{\rho_{11}}{\rho_{m11}})\right] \\ -\frac{1}{\delta_{21}}\left[v_{f21}\rho_{21}(1-\frac{\rho_{21}}{\rho_{m21}})-U\right] \\ -\frac{1}{\delta_{12}}\left[v_{f22}\rho_{22}(1-\frac{\rho_{22}}{\rho_{m22}})-v_{f21}\rho_{21}(1-\frac{\rho_{21}}{\rho_{m21}})\right] \end{bmatrix} \tag{48}$$

$$\mathbf{g}=\begin{bmatrix} U \\ 0 \\ -U \\ 0 \end{bmatrix} \tag{49}$$

Since we need to equate the travel times on alternate routes according to the UE DTR problem formulation presented in the previous section, we take the new transformed state variable y as the difference in travel times. Differentiating the equation representing y in terms of the state variables introduces the input split factor into the dynamic equation. Therefore, that transformed equation can be used to design the input that cancels the nonlinearities of the system and introduces a design input V, which can be used to place the poles of the error equation for asymptotic stability. These steps are shown below.

The variable y is equal to the difference in the travel time on the two sections.

$$y(t)=\left[\frac{d_{11}}{v_{f11}(1-\frac{\rho_{11}}{\rho_{m11}})}+\frac{d_{12}}{v_{f12}(1-\frac{\rho_{12}}{\rho_{m12}})}\right]-\left[\frac{d_{21}}{v_{f21}(1-\frac{\rho_{21}}{\rho_{m21}})}+\frac{d_{22}}{v_{f22}(1-\frac{\rho_{22}}{\rho_{m22}})}\right] \tag{50}$$

This equation can be differentiated with respect to time to give the travel time difference dynamics:

$$\dot{y}=\frac{k_1\dot{\rho}_{11}}{(k_2-\rho_{11})^2}+\frac{k_3\dot{\rho}_{12}}{(k_4-\rho_{12})^2}-\frac{k_5\dot{\rho}_{21}}{(k_6-\rho_{21})^2}-\frac{k_7\dot{\rho}_{22}}{(k_8-\rho_{22})^2} \tag{51}$$

By substituting Eqs. (40-43) in (51), we obtain

$$\dot{y}=-\frac{k_1\frac{1}{\delta_{11}}\left[v_{f11}\rho_{11}(1-\frac{\rho_{11}}{\rho_{m11}})-\beta U\right]}{(k_2-\rho_{11})^2}-\frac{k_3\frac{1}{\delta_{12}}\left[v_{f12}\rho_{12}(1-\frac{\rho_{12}}{\rho_{m12}})-v_{f11}\rho_{11}(1-\frac{\rho_{11}}{\rho_{m11}})\right]}{(k_4-\rho_{12})^2}+\frac{k_5\frac{1}{\delta_{21}}\left[v_{f21}\rho_{21}(1-\frac{\rho_{21}}{\rho_{m21}})+\beta U-U\right]}{(k_6-\rho_{21})^2}+\frac{k_7\frac{1}{\delta_{12}}\left[v_{f22}\rho_{22}(1-\frac{\rho_{22}}{\rho_{m22}})-v_{f21}\rho_{21}(1-\frac{\rho_{21}}{\rho_{m21}})\right]}{(k_8-\rho_{22})^2} \tag{52}$$

This equation can be rewritten in the following form:

$$\dot{y}=F+G\beta \tag{53}$$

where

$$F=-\frac{k_1\frac{1}{\delta_{11}}\left[v_{f11}\rho_{11}(1-\frac{\rho_{11}}{\rho_{m11}})\right]}{(k_2-\rho_{11})^2}-\tag{54}$$

$$\frac{k_3\frac{1}{\delta_{12}}\left[v_{f12}\rho_{12}(1-\frac{\rho_{12}}{\rho_{m12}})-v_{f11}\rho_{11}(1-\frac{\rho_{11}}{\rho_{m11}})\right]}{(k_4-\rho_{12})^2}+$$

$$\frac{k_5\frac{1}{\delta_{21}}\left[v_{f21}\rho_{21}(1-\frac{\rho_{21}}{\rho_{m21}})-U\right]}{(k_6-\rho_{21})^2}+$$

$$\frac{k_7\frac{1}{\delta_{22}}\left[v_{f22}\rho_{22}(1-\frac{\rho_{22}}{\rho_{m22}})-v_{f21}\rho_{21}(1-\frac{\rho_{21}}{\rho_{m21}})\right]}{(k_8-\rho_{22})^2}$$

$$G=[\frac{k_1\frac{1}{\delta_{11}}}{(k_2-\rho_{11})^2}+\frac{k_5\frac{1}{\delta_{21}}}{(k_6-\rho_{21})^2}]U\tag{55}$$

Hence, a feedback linearization control law similar to the one given by (21) can be designed to cancel the nonlinearities and provide the desired error dynamics. The law used is

$$\beta=G^{-1}(-F+v)\tag{56}$$

which gives the closed loop dynamics as

$$\dot{y}=v\tag{57}$$

The relative degree of the system is one, and the system order is four, and we need to test the stability or boundedness of the three internal states. Now our task is to obtain the other three independent state variables. Since this is a single input system, to obtain these state variables, we should satisfy $L_g\eta_k(\mathbf{x})=0$ as follows:

$$\frac{\partial\eta_i}{\partial\rho_{11}}=\frac{\partial\eta_i}{\partial\rho_{21}},\, i=1,2,3\tag{58}$$

$$\eta_1=\rho_{12}$$

$$\eta_2=\rho_{22}$$

$$\eta_3=\rho_{11}+\rho_{21}$$

The state variable $\eta$ is bounded since the densities on the sections cannot exceed the corresponding jam densities. We use the same argument as for the previous example for this claim.

$$\eta_1 \le \rho_{m12} \quad (59)$$
$$\eta_2 \le \rho_{m22}$$
$$\eta_3 \le \rho_{m11} + \rho_{m12}$$

and hence the overall system is exponentially stable ($y \to 0$) if we choose $v = -Ky, \quad K > 0$, and y asymptotically goes to zero as $y(t) = y(0)e^{-Kt}$. This implies that when a splitting value based on (56) is utilized, the difference in travel time of two alternate routes will go to zero at an exponential rate. Hence, the closed loop traffic system controlled by the proposed feedback linearization law is exponentially stable and has desired transient behavior.

# 6. Solution for The One-Origin, One-Destination Case With Multiple Routes With Multiple Sections

In this section, we give a generalized solution for the n alternate route DTR problem described in section 2. The space discretized flow equations used for the n alternate routes and n sections are given by (12) and (13). The number of sections for each alternate route i is denoted by $n_i$. We are considering full state observation, which is used for estimating (sensing) the travel times on the various alternate routes. The dynamics can be written as

$$\dot{\rho}_{ij} = \frac{1}{\delta_{ij}}\left[ v_{fij-1}\rho_{ij-1}(1 - \frac{\rho_{ij-1}}{\rho_{mij-1}}) - v_{fij}\rho_{ij}(1 - \frac{\rho_{ij}}{\rho_{mij}}) \right] \quad (60)$$

when

$$(i,j) = ((1,2),...,(1,n_1),(2,2),...,(2,n_2),...,(n,2),...,(n,n_n))$$

$$\dot{\rho}_{ij} = \frac{1}{\delta_{ij}}\left[ \beta_i U - v_{fij}\rho_{ij}(1 - \frac{\rho_{ij}}{\rho_{mij}}) \right] \quad (61)$$

when $(i,j) = ((1,1),(2,1)...(n,1))$

We have considered a simple first order travel time function, which is obtained by dividing the length of a section by the average velocity of vehicles on it. According to that, we approximate travel time for a route as

$$\chi_i(t) = \sum_{j=1}^{n_i} \frac{d_{ij}}{v_{fij}(1 - \frac{\rho_{ij}}{\rho_{mij}})} \quad (62)$$

The system can be written in the standard nonlinear input affine form:

$$\dot{\mathbf{x}}(t) = \mathbf{f}(\mathbf{x},t) + \mathbf{g}(\mathbf{x},t)\mathbf{u}(t)$$
$$\mathbf{y}(t) = \mathbf{h}(\mathbf{x},t) \quad (63)$$

where

$$\mathbf{x} = [\rho_{11}...\rho_{1n_1}\cdots\rho_{n1}\cdots\rho_{nn_n}]' \quad ,\mathbf{u}(t) = [\beta_1...\beta_{n-1}] \quad (64)$$

The output vector is denoted by y, and is given by:

$$y = [y_1\ y_2 \ldots y_i \ldots y_{n-1}] \tag{65}$$

where

$$y_i = \chi_{i+1}(t) - \chi_i(t) \tag{66}$$

This equation can be differentiated with respect to time to give the travel time difference dynamics.

$$\dot{y}_i = \sum_{j=1}^{n_{i+1}} \frac{k_{i+1j1}\dot{\rho}_{i+1j}}{k_{i+1j2}(1-\frac{\rho_{i+1j}}{\rho_{mi+1j}})^2} - \sum_{j=1}^{n_i} \frac{k_{ij1}\dot{\rho}_{ij}}{k_{ij2}(1-\frac{\rho_{ij}}{\rho_{mij}})^2} \tag{67}$$

where $k_{ijp}$ denotes a constant p=1 or 2 that belongs to section j of route i, similar to the constant k described for (30). The system (67) is in the form of (16) and can be represented in the form of (17) by determining the values of **A**(**x**) and **B**(**x**). The control law (21) provides us with user equilibrium for the DTR problem.

The input appears in all the output equations after differentiating them one time. Hence, the relative degree of the system is n-1. The order of the system is $\sum_{i=1}^{n} n_i$. Since the densities on the sections are bounded by jam density values, the independent state variables $\eta_i, i = 1,2,\ldots \sum_{i=1}^{n} n_i + 1 - n$ are also bounded.

## 7. Simulations

Several simulation studies are performed to demonstrate the utilization of the feedback linearization technique presented in this paper. The test network, which consists of two alternate routes, is shown in Figure 7.2. Three different simulation scenarios that were chosen are:

(1) Model without any parametric uncertainties and full user compliance
(2) Model with parametric uncertainties and full user compliance
(3) Model with parametric uncertainties and partial user compliance

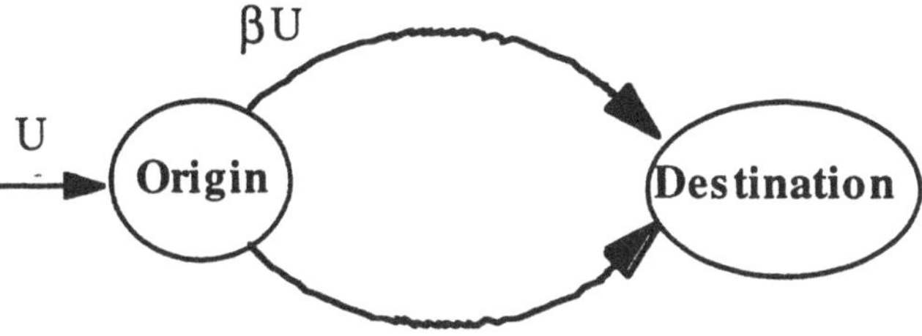


**Figure 7.2**: *Sample network.*

The input function is assumed to be a sinusoidal function, which reaches a predefined peak value and then settles at a constant value for the rest of the simulation period. This function emulates the peak hour demand that reaches its maximum value at a certain time, and then settles at a constant value when the peak period is over. In this specific simulation study, the peak period is assumed to be one hour.

*Scenario 1: Model with full user compliance and without any uncertainties*

In this scenario, the controller has perfect knowledge of the parameters of the traffic model. In addition, full compliance of the users to the diversion commands is assumed by the controller. The system dynamics model also simulates full compliance of the users to the controller's diversion commands. In this case, since the controller has the complete knowledge of the system dynamics, it is able to perform exact cancellation of the system nonlinearities and attain exponential error convergence. This result is shown in Figure 7.3 and Figure 7.4.

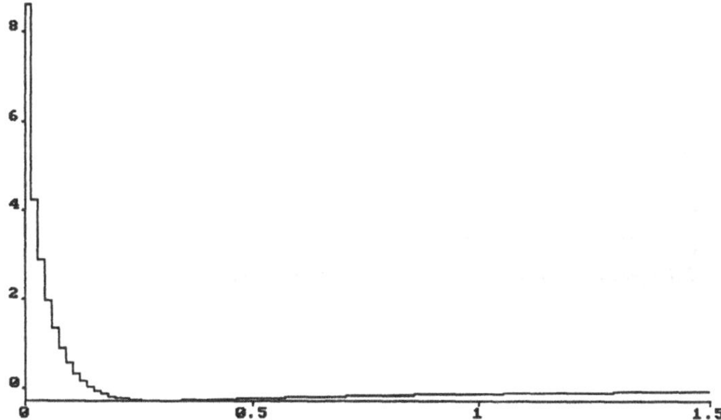


**Figure 7.3**: *Differences in travel times (secs) for scenario 1.*

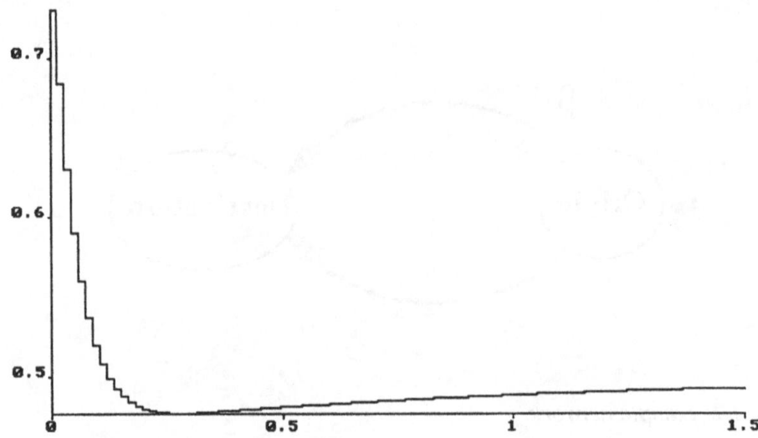


**Figure 7.4**: *Split factors for scenario 1.*

*Scenario 2: Model with full user compliance and uncertainties*

In this scenario, the controller does not have perfect knowledge of the parameters of the traffic model. In order to simulate the effects of uncertainties, ± 30% errors have been assumed. However, full compliance of the users to the diversion commands is assumed by the controller and is also simulated by the system model. In this case, since the controller does not have complete knowledge of the system dynamics, it is not able to perform exact cancellation of the system nonlinearities and attain exponential error convergence. However, as can be seen in Figure 7.5 and Figure 7.6, the results obtained by using this controller, even with such relatively large parametric uncertainty, are highly encouraging.

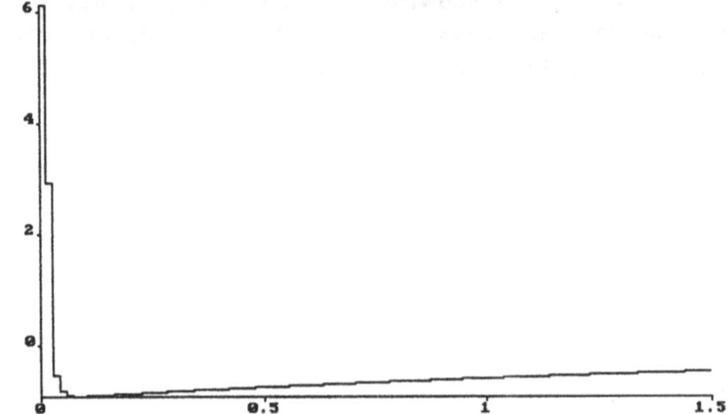


**Figure 7.5**: *Differences in travel times (secs) for scenario 2.*

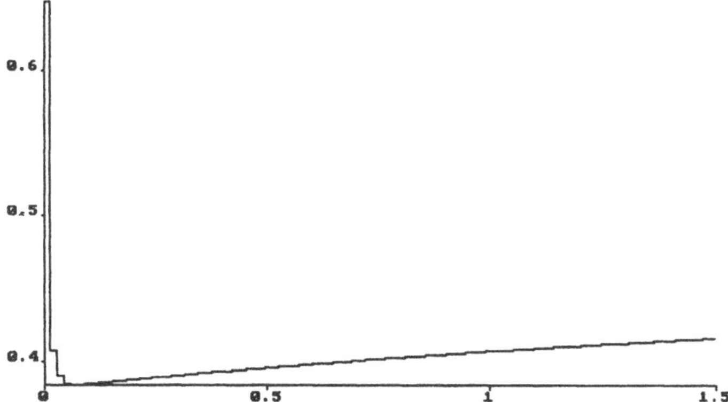


**Figure 7.6**: *Split factors for scenario 2.*

*Scenario 3: Model with partial user compliance and uncertainties*

In this scenario, we assume both partial user compliance (80%), and the existence of parametric uncertainty in the model. As can be seen in Figure 7.7 and Figure 7.8, the fluctuations of differences in travel are much higher than the previous scenarios, and it takes the controller a longer time to attain error convergence. However, even with partial user compliance and fairly large parametric uncertainties, the system stabilizes and the differences in travel times asymptotically converge at a desirable rate.

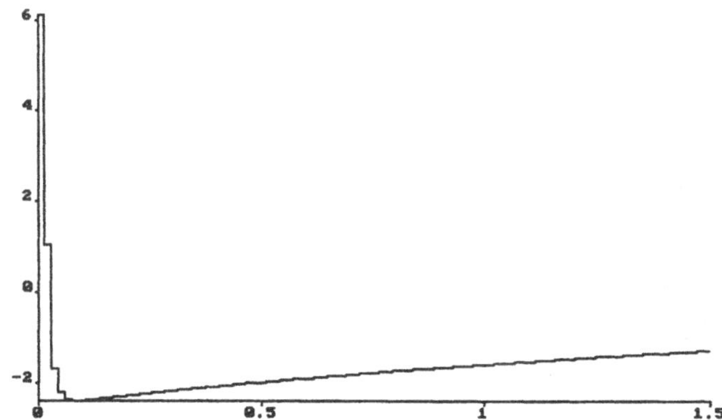


**Figure 7.7**: *Differences in travel times (secs) for scenario 3.*

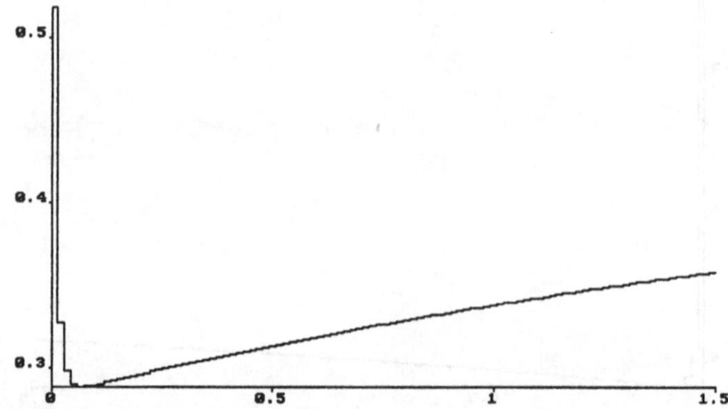


**Figure 7.8**: *Split factors for scenario 3.*

*Scenario 4: Model with partial user compliance and uncertainties, and a linear PI controller*

In this scenario, we assume both partial user compliance (80%), and the existence of parametric uncertainty in the model, and we use a linear PI (Proportional, Integral) controller. As is evident here, the performance of this controller is inferior to the feedback linearization controller (Figure7.9 and Figure 7.10).

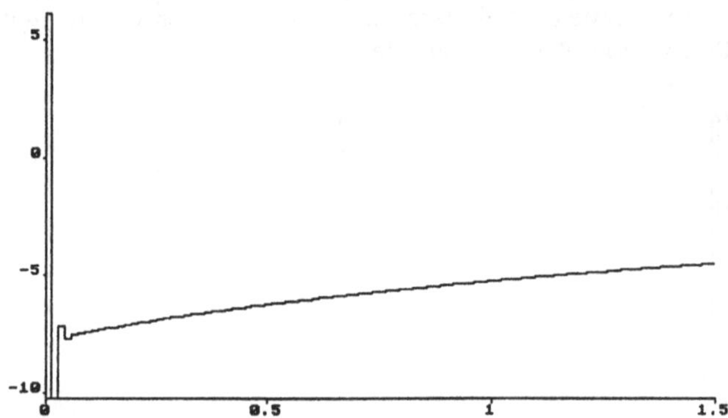


**Figure 7.9**: *Differences in travel times (secs) for scenario 4.*

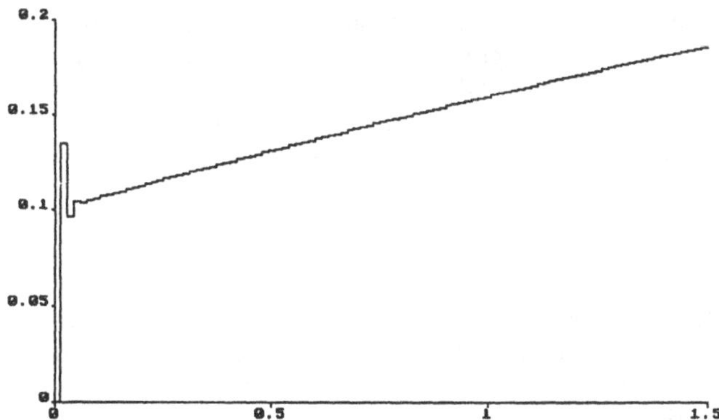


**Figure 7.10**: *Split factors for scenario 4.*

*Scenario 5: Model with partial user compliance, uncertainties, and dynamic velocity relationship*

In this scenario, we assume both partial user compliance (80%), and the existence of parametric uncertainty in the model. We also consider a dynamic relationship for velocity to represent the shock wave dynamics in the model. The results of using the feedback linearization to uncertainties obtained by under-modeling of this kind are also satisfactory (Figure7.11 and Figure 7.12).

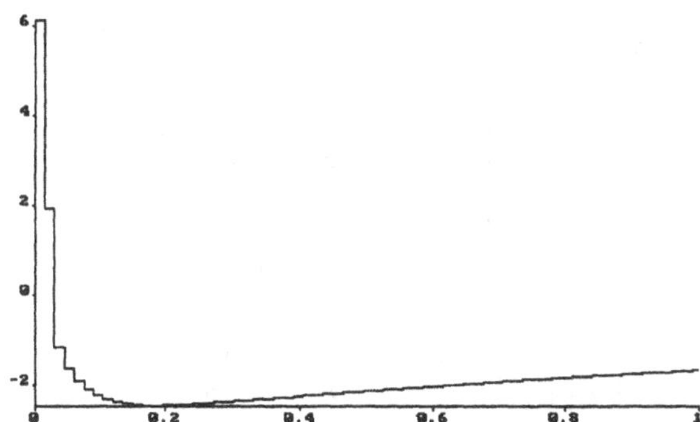


**Figure7.11**: *Differences in travel times (secs) for scenario 5.*

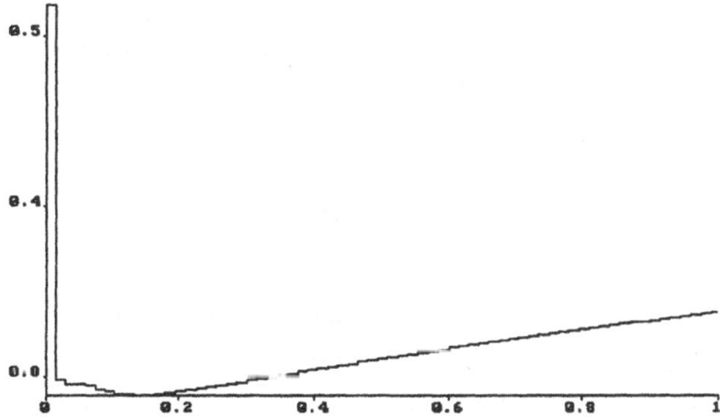


**Figure7.12**: *Split factors for scenario 5.*

In all the scenarios, there is no congestion created in the two routes. In general, the control algorithm cannot prevent congestion if there is a very large inflow of traffic. Consider the situation where the inflow traffic is so high that it produces traffic, which is greater than the overall capacity of the two routes; then the split factor control cannot avoid congestion. In that case, some traffic from the inflow itself would have to be diverted.

**Simulation environment**

The simulation environment we used is SIMNON [5], which is a special programming language developed in the Lund Institute of Technology, Sweden, for simulating dynamic systems described as ordinary differential equations, as difference equations, or as combinations of both. This program is available in DOS and Windows environments.

# 8. Sliding Mode Control for Point Diversion

We can also analyze and design control when there are uncertainties in the system, which is bound to be happen in practice. When there are uncertainties in the system, the control law (21) can not be directly utilized, but it will have to be modified. Sliding mode control design is a useful technique to deal with uncertainties of systems in canonical form.

**Sample Problem (Two alternate routes with one section)**

In order to illustrate the ideas discussed above, we have designed a feedback control law for two alternate routes problem with single section each. The space discretized flow equations used for the two alternate routes are:

$$\dot{\rho}_1 = -\frac{1}{\delta_1}\left[ v_{f1}\rho_1(1-\frac{\rho_1}{\rho_{m1}}) - \beta u \right] + \omega_1, \tag{68}$$

$$\dot{\rho}_2 = -\frac{1}{\delta_2}\left[ v_{f2}\rho_2(1-\frac{\rho_2}{\rho_{m2}}) + \beta u - u \right] + \omega_2, \tag{69}$$

where $\omega_1$ and $\omega_2$ are the terms representing uncertainties of the system equations. We have considered a simple first order travel time function, which is obtained by dividing the length of a section by average velocity of vehicles on it. According to that, the travel time can be calculated as

$$\chi_1(k) = d_1 / [v_{f1}(1-\frac{\rho_1}{\rho_{m1}})], \tag{70}$$

$$\chi_2(k) = d_2 / [v_{f2}(1-\frac{\rho_2}{\rho_{m2}})], \tag{71}$$

where, $d_1$ and $d_2$ are section lengths, $v_{f1}$ and $v_{f2}$ are the free flow speeds of each section, and $\rho_{m1}$ and $\rho_{m2}$ are the maximum (jam) densities of each section. Since we need to equate the travel times in the two freeways, we take the new transformed state variable y as the difference in travel times. Differentiating the equation representing y in terms of the state variables introduces the input split factor into the dynamic equation.

The variable y is equal to the difference in the travel time on the two sections.

$$y = \frac{k_1}{(k_2 - \rho_1)} - \frac{k_3}{(k_4 - \rho_2)} \tag{72}$$

where $k_1 = \frac{d_1 . \rho_{m1}}{v_{f1}}, k_2 = \rho_{m1}, k_3 = \frac{d_2 . \rho_{m2}}{v_{f2}}, k_4 = \rho_{m2}$.

This equation can be differentiated with respect to time to give the travel time difference dynamics.

$$\dot{y} = \frac{k_1 \dot{\rho}_1}{(k_2 - \rho_1)^2} - \frac{k_3 \dot{\rho}_2}{(k_4 - \rho_2)^2} \tag{73}$$

By substituting (68) and (69) in (73) we obtain

$$\dot{y} = -\frac{k_1\left( v_{f1}\rho_1(1 - \frac{\rho_1}{\rho_{m1}}) - \beta u + \delta_1 \omega_1 \right)}{\delta_1 (k_2 - \rho_1)^2} + \frac{k_3\left( v_{f2}\rho_2(1 - \frac{\rho_2}{\rho_{m2}}) + \beta u - u + \delta_2 \omega_2 \right)}{\delta_2 (k_4 - \rho_2)^2} \tag{74}$$

This equation can be rewritten in the following form.

$$\dot{y} = F + G\beta + f \tag{75}$$

where

$$F = -\frac{k_1 v_{f1} \rho_1}{\delta_1 (k_2 - \rho_1)^2}(1 - \frac{\rho_1}{\rho_{m1}}) + \frac{k_3}{\delta_2 (k_4 - \rho_2)^2}\left( (1 - \frac{\rho_2}{\rho_{m2}}) v_{f2} \rho_2 - u \right) \tag{76}$$

$$G = \left( \frac{k_1}{\delta_1 (k_2 - \rho_1)^2} + \frac{k_3}{\delta_2 (k_4 - \rho_2)^2} \right) u \tag{77}$$

$$f = \frac{k_1 \omega_1}{(k_2 - \rho_1)^2} - \frac{k_3 \omega_2}{(k_4 - \rho_2)^2} \tag{78}$$

Hence a feedback linearization control law can be designed to cancel the non-linearities and provide the desired error dynamics. The sliding mode control law used is

$$\beta = G^{-1}\left(-F - \hat{f} - k\,\mathrm{sgn}(y)\right), \quad k \geq \phi \tag{79}$$

where $\phi$ is a known function which gives the error estimate bound on f. If there were explicit uncertainties on G, then we would use control law similar to (47) and (48) in chapter 4.

**Sample Problem (Two alternate routes with two sections)**

Now, we extend the above problem to two sections case and follow the same steps for designing a new controller for this extended system. The space discretized flow equations used for the two alternate routes are:

$$\dot{\rho}_{11} = -\frac{1}{\delta_{11}}\left[v_{f11}\rho_{11}(1-\frac{\rho_{11}}{\rho_{m11}}) - \beta u\right] + \omega_{11}, \tag{80}$$

$$\dot{\rho}_{12} = -\frac{1}{\delta_{12}}\left[v_{f12}\rho_{12}(1-\frac{\rho_{12}}{\rho_{m12}}) - v_{f11}\rho_{11}(1-\frac{\rho_{11}}{\rho_{m11}})\right] + \omega_{12}, \tag{81}$$

$$\dot{\rho}_{21} = -\frac{1}{\delta_{21}}\left[v_{f21}\rho_{21}(1-\frac{\rho_{21}}{\rho_{m21}}) + \beta u - u\right] + \omega_{21} \tag{82}$$

$$\dot{\rho}_{22} = -\frac{1}{\delta_{12}}\left[v_{f22}\rho_{22}(1-\frac{\rho_{22}}{\rho_{m22}}) - v_{f21}\rho_{21}(1-\frac{\rho_{21}}{\rho_{m21}})\right] + \omega_{22} \tag{83}$$

The travel times are:

$$\chi_1(t) = \frac{d_{11}}{v_{f11}(1-\frac{\rho_{11}}{\rho_{m11}})} + \frac{d_{12}}{v_{f12}(1-\frac{\rho_{12}}{\rho_{m12}})}, \tag{84}$$

$$\chi_2(t) = \frac{d_{21}}{v_{f21}(1-\frac{\rho_{21}}{\rho_{m21}})} + \frac{d_{22}}{v_{f22}(1-\frac{\rho_{22}}{\rho_{m22}})}, \tag{85}$$

where, $d_1$ and $d_2$ are section lengths, $v_{f1}$ and $v_{f2}$ are the free flow speeds of each section, and $\rho_{m1}$ and $\rho_{m2}$ are the maximum (jam) densities of each section.

We take the new transformed state variable y as the difference in travel times. Differentiating the equation representing y in terms of the state variables introduces the input split factor into the dynamic equation.

The variable y is equal to the difference in the travel time on the two sections.

$$y(t) = [\frac{d_{11}}{v_{f11}(1-\frac{\rho_{11}}{\rho_{m11}})} + \frac{d_{12}}{v_{f12}(1-\frac{\rho_{12}}{\rho_{m12}})}] - [\frac{d_{21}}{v_{f21}(1-\frac{\rho_{21}}{\rho_{m21}})} + \frac{d_{22}}{v_{f22}(1-\frac{\rho_{22}}{\rho_{m22}})}] \tag{86}$$

This equation can be differentiated with respect to time to give the travel time difference dynamics.

$$\dot{y}=\frac{k_1\dot{\rho}_{11}}{(k_2-\rho_{11})^2}+\frac{k_3\dot{\rho}_{12}}{(k_4-\rho_{12})^2}-\frac{k_5\dot{\rho}_{21}}{(k_6-\rho_{21})^2}-\frac{k_7\dot{\rho}_{22}}{(k_8-\rho_{22})^2} \tag{87}$$

By substituting (73)- (76) in (84) we obtain

$$\dot{y}=-\frac{k_1\frac{1}{\delta_{11}}\left[v_{f11}\rho_{11}(1-\frac{\rho_{11}}{\rho_{m11}})-\beta u\right]+k_1\omega_{11}}{(k_2-\rho_{11})^2}- \tag{88}$$

$$\frac{k_3\frac{1}{\delta_{12}}\left[v_{f12}\rho_{12}(1-\frac{\rho_{12}}{\rho_{m12}})-v_{f11}\rho_{11}(1-\frac{\rho_{11}}{\rho_{m11}})\right]+k_3\omega_{12}}{(k_4-\rho_{12})^2}+$$

$$\frac{k_5\frac{1}{\delta_{21}}\left[v_{f21}\rho_{21}(1-\frac{\rho_{21}}{\rho_{m21}})+\beta u-u\right]+k_5\omega_{21}}{(k_6-\rho_{21})^2}+$$

$$\frac{k_7\frac{1}{\delta_{22}}\left[v_{f22}\rho_{22}(1-\frac{\rho_{22}}{\rho_{m22}})-v_{f21}\rho_{21}(1-\frac{\rho_{21}}{\rho_{m21}})\right]+k_7\omega_{22}}{(k_8-\rho_{22})^2}$$

This equation can be rewritten in the following form.

$$\dot{y}=F+G\beta+f \tag{89}$$

where

$$F=-\frac{k_1\frac{1}{\delta_{11}}\left[v_{f11}\rho_{11}(1-\frac{\rho_{11}}{\rho_{m11}})\right]}{(k_2-\rho_{11})^2}-\frac{k_3\frac{1}{\delta_{12}}\left[v_{f12}\rho_{12}(1-\frac{\rho_{12}}{\rho_{m12}})-v_{f11}\rho_{11}(1-\frac{\rho_{11}}{\rho_{m11}})\right]}{(k_4-\rho_{12})^2}+\frac{k_5\frac{1}{\delta_{21}}\left[v_{f21}\rho_{21}(1-\frac{\rho_{21}}{\rho_{m21}})-u\right]}{(k_6-\rho_{21})^2}+\frac{k_7\frac{1}{\delta_{22}}\left[v_{f22}\rho_{22}(1-\frac{\rho_{22}}{\rho_{m22}})-v_{f21}\rho_{21}(1-\frac{\rho_{21}}{\rho_{m21}})\right]}{(k_8-\rho_{22})^2} \tag{90}$$

$$G=[\frac{k_1\frac{1}{\delta_{11}}}{(k_2-\rho_{11})^2}+\frac{k_5\frac{1}{\delta_{21}}}{(k_6-\rho_{21})^2}]u \tag{91}$$

$$f=\frac{k_1\omega_{11}}{(k_2-\rho_{11})^2}-\frac{k_3\omega_{12}}{(k_4-\rho_{12})^2}+\frac{k_5\omega_{21}}{(k_6-\rho_{21})^2}+\frac{k_7\omega_{22}}{(k_8-\rho_{22})^2}$$

The control law for this case is

$$\beta=G^{-1}\left(-F-\hat{f}-k\,\mathrm{sgn}(y)\right),\quad k\geq\phi \tag{92}$$

which guarantees convergent travel time difference.

### Solution for the Generalized DTR Problem for Multiple Routes with Multiple Sections

In this section, we give a generalize solution for the n alternate route DTR problem. Number of sections for each alternate route i is denoted by $n_i$. We are considering full state observation, which is used for estimating (sensing) the travel times on the various alternate routes. The dynamics can be written as

$$\dot{\rho}_{ij}=\frac{1}{\delta_{ij}}\left[v_{fij-1}\rho_{ij-1}(1-\frac{\rho_{ij-1}}{\rho_{mij-1}})-v_{fij}\rho_{ij}(1-\frac{\rho_{ij}}{\rho_{mij}})\right]+\omega_{ij} \tag{93}$$

when

$$(i,j)=((1,2),...,(1,n_1),(2,2),...,(2,n_2),...,(n,2),...,(n,n_n))$$

$$\dot{\rho}_{ij} = \frac{1}{\delta_{ij}}\left[\beta_i u - v_{fij}\rho_{ij}(1 - \frac{\rho_{ij}}{\rho_{mij}})\right] + \omega_{ij} \tag{94}$$

$$\text{when } (i,j) = ((1,1),(2,1)...(n,1))$$

We have considered a simple first order travel time function, which is obtained by dividing the length of a section by average velocity of vehicles on it. According to that, we approximate travel time for a route as

$$\chi_i(t) = \sum_{j=1}^{n_i} \frac{d_{ij}}{v_{fij}(1 - \frac{\rho_{ij}}{\rho_{mij}})} \tag{95}$$

The system can be written in the standard nonlinear input affine form

$$\begin{aligned} \dot{\mathbf{x}}(t) &= \mathbf{f}(\mathbf{x},t) + \mathbf{g}(\mathbf{x},t)\mathbf{u}(t) \\ \mathbf{y}(t) &= \mathbf{h}(\mathbf{x},t) \end{aligned} \tag{96}$$

where

$$\mathbf{x} = [\rho_{11} \cdots \rho_{1n_1} \cdots \rho_{n1} \cdots \rho_{nn_n}]' \quad , \mathbf{u}(t) = [\beta_1 ... \beta_{n-1}] \tag{97}$$

The output vector is denoted by y, and is given by:

$$\mathbf{y} = [y_1 \ y_2 \cdots y_i \cdots y_{n-1}] \tag{98}$$

where,

$$y_i = \chi_{i+1}(t) - \chi_i(t) \tag{99}$$

This equation can be differentiated with respect to time to give the travel time difference dynamics.

$$\dot{y}_i = \sum_{j=1}^{n_{i+1}} \frac{k_{i+1j1}\dot{\rho}_{i+1j}}{k_{i+1j2}(1 - \frac{\rho_{i+1j}}{\rho_{mi+1j}})^2} - \sum_{j=1}^{n_i} \frac{k_{ij1}\dot{\rho}_{ij}}{k_{ij2}(1 - \frac{\rho_{ij}}{\rho_{mij}})^2} \tag{100}$$

where $k_{ijp}$ denotes a constant p=1 or 2 that belongs to section j of route i, similar to the constant k described for (72). This system is input-output square and can be solved since each output equation is decoupled and can be solved by the sliding mode control law.

**Simulation**

The test network consists of two alternate routes. The demand function is a constant flow of 325 vehicles per 15 minutes and it is then increased to 1000 vehicles per fifteen minutes. We have assumed full compliance of users for this sample case. The results shown in the figure below illustrate that the travel times become equal after the successful implementation of the controller developed in this paper. One drawback which is evident from the figure is the high frequency chattering encountered using the sliding mode control. To eliminate this problem, we can use the boundary layer method to replace a continuous controller inside the boundary layer as described in detail in chapter 4.

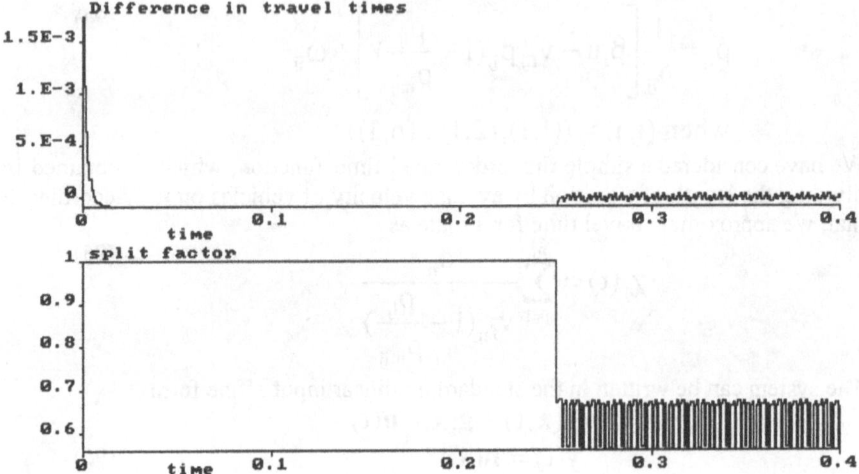


**Figure7.13**: *Simulation Results using Sliding Mode Control*

# 9. Summary

In this chapter, we have addressed the real-time traffic control problem for point diversion. A feedback model is developed for control purposes, and feedback linearization technique is used to design this feedback controller. To counter partially known disturbances, sliding mode control design is also shown. First, the simplest case, with two alternate routes consisting of a single section each, is studied and a feedback controller using feedback linearization technique is developed. Second, the case with two alternate routes with two discrete sections is analyzed, and a feedback controller using feedback linearization technique is also developed. Finally, the general case with multiple alternate routes divided into multiple sections is analyzed, and a general solution is proposed. To illustrate the above models, simulation runs are performed for three different scenarios for a network topology of two alternate routes. The feedback controller developed for this test network performed fairly well for all three scenarios. An important finding of this simulation study was the robustness of the controller even for situations where parametric uncertainties exist and partial user compliance is employed. Therefore, we can conclude that feedback linearization is an effective method for designing real-time traffic control systems. We also designed sliding mode controllers for the different sub-problems.

# 10. Exercises

### Questions

Question 1:What knowledge of the system is required for the controller in feedback linearization?
Question 2: What is the canonical form of a system?.
Question 3: What can happen when the controller knowledge about the system is not correct in feedback linearization?
Question 4: What kind of uncertainties can sliding mode control handle?
Question 5: What is the price you pay for using sliding mode control?
Question 6: How can one deal with the chattering problem in sliding mode control?
Question 7: How can one obtain the canonical form for the dynamic traffic routing problem for point diversion?
Question 8: Compare feedback linearization and sliding mode control especially as they apply to the point diversion problem.

### Problems

*Problem 1*

Derive feedback linearization control law for two alternate routes with one section each. Perform simulations similar to the ones shown in the chapter. Now assume that f is less than or equal to 20% of F in sliding mode control. Based on this, derive the sliding mode control law for the problem, and perform simulations.

*Problem 2*

Derive feedback linearization control law for two alternate routes with two sections each. Perform simulations similar to the ones shown in the chapter. Now assume that f is less than or equal to 20% of F in sliding mode control. Based on this, derive the sliding mode control law for the problem, and perform simulations.

*Problem 3*

Derive feedback linearization control law for three alternate routes with one section each. Perform simulations similar to the ones shown in the chapter. Now assume that f is less than or equal to 20% of F in sliding mode control. Based on this, derive the sliding mode control law for the problem, and perform simulations.

*Problem 4*

Derive feedback linearization control law for three alternate routes with two sections each. Perform simulations similar to the ones shown in the chapter. Now assume that f is less than or equal to 20% of F in sliding mode control. Based on this, derive the sliding mode control law for the problem, and perform simulations.

## 11. References

1. Papageorgiou M. and Messmer A. (1991) Dynamic network traffic assignment and route guidance via feedback regulation. Transportation Research Board, Washington, D.C..
2. Isidori A. (1989) *Nonlinear Control Systems*. Springer-Verlag.
3. Slotine J. J. E. and Li W. (1991) *Applied Nonlinear Control*. Prentice Hall.
4. Godbole D. N. and Sastry S. S. (1995) Approximate decoupling and asymptotic tracking for MIMO systems. *IEEE Transactions on Automatic Control*. **40**, 441-450
5. Elmqvist H. (1975) SIMNON - An Interactive Simulation Program for Nonlinear Systems. Dept. of Automatic Control, Lund University of Lund, Sweden, Report TFRT-3091.

# CHAPTER 8
# FEEDBACK CONTROL FOR NETWORK LEVEL DYNAMIC TRAFFIC ROUTING

## Objectives

- To develop models for network level traffic systems.
- To design feedback controllers for network level traffic problems in user-equilibrium as well as system optimal settings.

## 1. Introduction

The technique we propose in this chapter solves the network-wide system-optimal and user-equilibrium DTA/DTR problem using real-time feedback control. We employ nonlinear $H_\infty$ feedback control design methodology to produce the solution of the problem, which also provides robustness against bounded disturbances. The nonlinear $H_\infty$ problem is seen as a two player zero-sum differential game played by the control action (route guidance system) and the disturbances in the system (bounded unmodeled dynamics and uncertainties) [1-8]. The solution of the nonlinear $H_\infty$ problem relies on solving a stationary Hamilton-Jacobi inequality [1-8]. The modeling paradigm of nonlinear $H_\infty$ approach is an exact match with the requirements of a network-wide DTA/DTR problem applicable to Advanced Traffic Management/Information Systems (ATMIS) of Intelligent Transportation Systems (ITS), because it solves the optimal dynamic routing problem by only performing simple algebraic operations in real-time, unlike existing techniques which rely on lengthy off-line/on-line mathematical operations. The theory developed for network-wide problem is applied to a sample network.

## 2. System Description

In this section, we present a mathematical model for the traffic system, which is in the form usable for the design of DTA/DTR feedback control. Many models have been proposed before, but the most appropriate model has been proposed by Papageorgiou [9]. In this section, we will present the same model with minor changes, which then will be used for feedback control design.

**System Network**

Following the general notation and development in [9], let N be the set of all the network nodes for the problem, M the set of all the networked links, $I_n$ the set of links entering node n, $O_n$ the set of links leaving node n, O the set of origin nodes, and D the set of destination nodes. Let $d_{ij}$ denote the origin-destination demands. where $i \in O$ and $j \in D$. Traffic flow entering a link m is shown by

$q_m$ and that leaving the same link by $Q_m$. Let $S^n$ denote the set of destination nodes which are reachable from a node n. There are $\ell_{nj}$ alternate routes from node n to destination node j, and $L^z_{nj}, z = 1,2,...,\ell_{nj}$ are the ordered sets of links included in alternate routes. $\Lambda_{nj}$ is the set of output links which connect the node to the destination j using one of the alternate $\ell_{nj}$ routes. $q^m_{nj}$, $m \in \Lambda_{nj}$ is the flow in the link m belonging to one of the routes for destination j flowing out from node n. The node variables are shown in Figure 8.1. The sum of all $q^m_{nj}$, $m \in \Lambda_{nj}$ for a node n for destination j is given by $q_{nj}$.

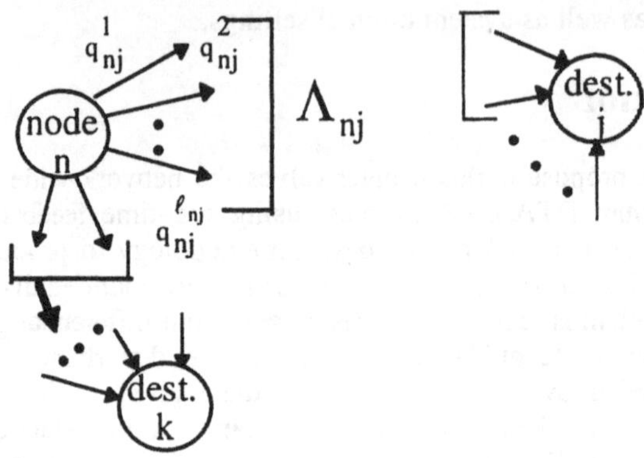


**Figure 8.1**: *Node Model*

There are two kinds of split factors, which can be used in system dynamics. One is destination based splits, which are given as ratio of the destination based flow on a link out of a node n and the total flow from the same node to the same destination. This is shown in equation (1). The constraint associated with this split factor inputs is given by (2).

$$\beta^m_{nj} = \frac{q^m_{nj}}{q_{nj}}, \quad m \in \Lambda_{nj} \tag{1}$$

$$\sum_{m \in \Lambda_{nj}} \beta^m_{nj} = 1 \tag{2}$$

Second kind of split factor input takes the ratio of the flow into a link m from a node n and the total flow from the same node, as shown in (3). The associated constraint is shown in (4).

$$\beta^m_n = \frac{q^m_n}{q_n}, \quad m \in \Lambda_n \tag{3}$$

$$\sum_{m \in \Lambda_n} \beta^m_n = 1 \tag{4}$$

The total number of independent split factor input variables, whether they are destination based or node based, is reduced by one because of the constraints (2) and (4).

Link variables can also be modeled based on either destination or independent of those. Let $S_m$ be the set of destination nodes, which are reachable through link m. The inflow into a link m is given by

$$q_m = \sum_{j \in S_m} \beta_{nj}^m q_{nj} \quad n \in N, m \in O_n \tag{5}$$

Let $q_{nj}$ denote the total flow leaving a node n for destination j. The composition rates on a link are given by (6) and the corresponding constraints are given by (7).

$$\gamma_{mj} = \beta_{nj}^m \frac{q_{nj}}{q_m}, \quad n \in N, m \in O_n, j \in S_m \tag{6}$$

$$\sum_{j \in S_m} \gamma_{mj} = 1 \tag{7}$$

The destination based link flow is given by

$$q_{nj} = \sum_{m \in I_n} Q_m \Gamma_{mj} + d_{nj} \quad n \in N, j \in S^n \tag{8}$$

where $\Gamma_{mj}$ is the fraction of traffic volume exiting a link m destined for destination j. The link model is illustrated in Figure 8.2.

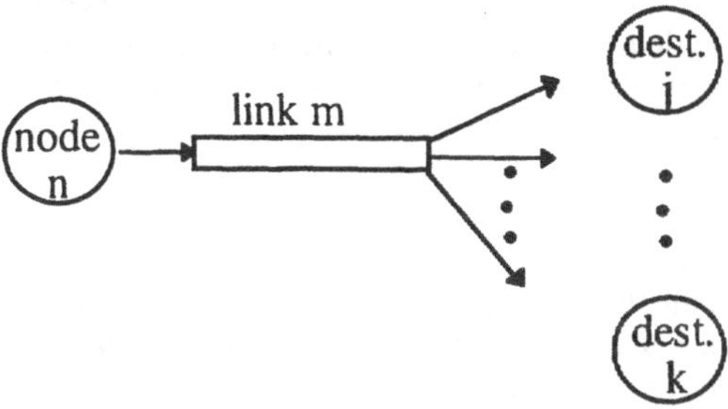


**Figure 8.2**: *Link Model*

**System Dynamics**

There are essentially two kinds of system dynamics which have been modeled for this problem : link based model, and route based model.

*Link Based Model*

In this scheme, the state variables are link density and composition rates. The link density dynamics are obtained from conservation equation, and are given by (9). The relationship between outflow and link density is shown by (10). There have been many other relationships shown in literature instead of (10).

$$\dot{\rho}_m(t) = \frac{1}{\delta_m}[q_m(t) - Q_m(t)] \tag{9}$$

$$Q_m(t) = q_{max,m}[1 - e^{-\rho_m(t)/\kappa_m}] \tag{10}$$

The composition rate dynamics can be represented as a time delay, where the amount of time delay is related to the travel time on the link. This dynamic relation is shown as

$$\Gamma_{mj}(t) = \gamma_{mj}(t - \chi_{mj}) \tag{11}$$

An alternate method models the dynamics of composition rate propagation as a first order filter, given by

$$\dot{\Gamma}_{mj}(t) = \alpha_m[\gamma_{mj} - \Gamma_{mj}] + \Gamma_{mj} \tag{12}$$

where $\alpha_m$ is either constant or a function of travel time. It is easier to use (12) for designing feedback control models. Using (11) instead of (12) makes the system infinite dimensional, and hence its control design becomes more difficult, especially when the delay also is state dependent. The time delay model assumes that the input composition rate into a link, travels without getting changed through the link with the travel speed and comes out of the link as the output composition rate. Which means that by shifting the plot of incoming composition rate of a link by the amount of travel time (if it is constant), we will obtain the outgoing composition rate of the same link. In the filter model also there is a delay associated with the composition rate propagation, but the filter also affects the behavior of the rates. This effect is seen in the transient behavior such as by observing the step or impulse response of the filter. The transient response indicators are usually response time, settling time, percent overshoot, etc. It seems that in real traffic conditions there should be some modification of the composition rates through a link, and that is why we consider the filter model to be more representative of the actual system behavior.

*Route Based Model*

In the route based model, the state variables are the destination based densities on the links of the system. The dynamic equations are shown in (13). The exiting composition rates are given by the ratio of the destination based density to the total link density as shown in (14).

$$\dot{\rho}_{mj}(t) = \frac{1}{\delta_m}[\gamma_{mj}(t)q_m(t) - \Gamma_{mj}(t)Q_m(t)] \tag{13}$$

$$\Gamma_{mj}(t) = \rho_{mj}(t)/\rho_m(t) \tag{14}$$

# 3. Dynamic Traffic Assignment Problem

The link based and route based models described in section 2.2 can be represented in the following form:

$$\dot{\mathbf{x}} = \mathbf{f}(\mathbf{x}) + \mathbf{g}(\mathbf{x})\mathbf{u} \tag{15}$$

For instance, a link based model with filter composition rate dynamics (12) and destination based split factors can be written in form (15) by using (1), (2), (5-10) and (12). A more complete form of model (15) would also represent disturbances, which come from uncertainties of the system. This new form is given by

$$\dot{\mathbf{x}} = \mathbf{f}(\mathbf{x}) + \mathbf{g}(\mathbf{x})\mathbf{u} + \mathbf{a}(\mathbf{x})\mathbf{w} \tag{16}$$

where w is the disturbance to the system model and a(x) the corresponding gain.

Consider the following representation of system, which is also shown in the Figure 8.3. In the figure, K represents the controller block and G the system block. First we will model the system for a nonlinear $H_\infty$ control design, which is the solution of a zero-sum, two player differential game with u(t) being one player and disturbance w(t) being the other.

$$G:\left\{\begin{array}{c} \dot{\mathbf{x}} = \mathbf{f}(\mathbf{x}) + \mathbf{g}(\mathbf{x})\mathbf{u} + \mathbf{a}(\mathbf{x})\mathbf{w} \\ \mathbf{y} = \mathbf{x} \\ \mathbf{z} = \begin{bmatrix} \mathrm{h}(\mathbf{x}) \\ \mathbf{u} \end{bmatrix} \end{array}\right\} \tag{17}$$

The function h(x) is very important in the network-wide user-equilibrium optimal dynamic traffic assignment control. The aim of the assignment is to reach user equilibrium by keeping cost on alternate routes the same. We can achieve that by taking the cost as the sum of the squares of the differences between the costs on the alternate routes. Then the controller will try to minimize the travel time keeping the controller cost low. In the actuation sense, which could be variable message sign, or direct communication with the vehicles, it makes sense to keep the variation of the split factors low with respect to time, so that the signs or commands don't change too rapidly [9]. In order to accomplish that, we add more states to the system which consist of the split factors also, and the control input becomes the derivative of the split factor values.

Note that due to the presence of origin-destination (OD) flows in the dynamics, the system dynamics are non-linear-time varying (NLTV). In order for us to use a stationary solution of the Hamilton-Jacobi (HJ) equation, we need to have a time invariant system. We can convert the NLTV system into a nonlinear-time-invariant (NLTI) system by introducing additional dynamic equations for OD flows using the assumptions from (13, 14), which consider an autonomous dynamic behavior of the OD flows. This extends the state variable vector by the additional OD flow dynamics.

Now, in order to formulate the problem in state feedback $H_\infty$ control, we need to identify z(t). For a system user-equilibrium, we can take z(t) as a cost function of the state variables with weights given to the function of states as well as to the variation of the split factors. That would in effect provide bounded variation of the split factor commands which is crucial for the actual implementation and effectiveness of the system. This further increases the size of the state variable vector by the number of split factor variables.

The DTA/DTR can be further solved in two ways depending on what we use for split factors i.e. destination based split factors or node based split factors. There are many subtle and apparent theoretical as well as practical implications of this choice. The theoretical implications are related to the controllability aspect when deciding on which split factor to use. Without any detailed analysis, it seems intuitive that the destination based split factor formulation will give a more controllable system dynamics than the node based split factors. However, the actual implementation of destination based split factors is not trivial. At present, VMS systems or other actuation methods can be used for node based splitting, and they would have to be either modified or designed in such a way that destination based splitting information can be provided to the drivers. In automated highway systems, or in general in a transportation system, where communication infrastructure is already present for infrastructure to vehicle communication (such as with in-vehicle route guidance system), the destination based split factors could be easily implemented, and be highly effective.

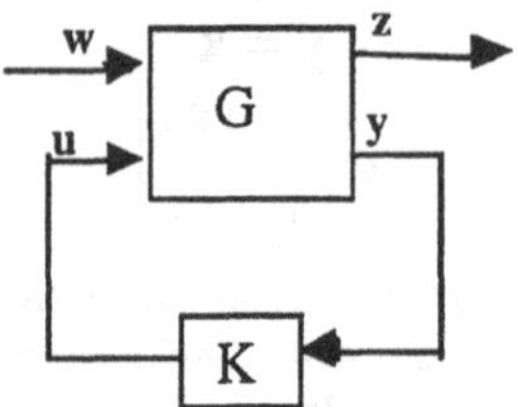


**Figure 8.3**: *Block diagram for nonlinear $H_\infty$ formulation*

Definition: System G/K is said to have $L_2$ gain less than or equal to _ for some _ > 0 if

$$\int_0^T \|z(t)\|^2 dt \le \gamma^2 \int_0^T \|w(t)\|^2 dt \qquad (18)$$

$\forall$ T > 0 and d(t) $\in L_2[0, T]$.

Nonlinear $H_\infty$ Control Problem: Find an output feedback controller K if any, such that the closed loop system $\Omega(G, K)$ is asymptotically stable and has $L_2$-gain ≤ __ $\forall$ T $\in R^+$.

Solution: The solution to this problem can be derived from the theory of dissipative systems [6], which also has implications to the theory of differential games [5]. In order to develop the solution, define a storage function as in [4] as:

$$V_s(\mathbf{x}) = \sup_{w \in L_2[0,T], x(o)=x} \frac{1}{2}\int_0^T (\|z(t)\|^2 - \gamma^2\|w(t)\|^2)dt \qquad (19)$$

Condition (18) is equivalent to $V_s(x) < \infty$, which in turn is true if and only if there exists a solution to the following integral dissipation inequality.

$$V(\mathbf{x}(t_1)) - V(\mathbf{x}(t_0)) \le \frac{1}{2}\int_{t_0}^{t_1}(\gamma^2\|\mathbf{w}(t)\|^2 - \|\mathbf{z}(t)\|^2)dt,\ V(0) = 0 \tag{20}$$

$\forall t_1 > t_0$ and $w \in L_2[t_0, t_1]$. Function $V(\mathbf{x})$ which satisfies (20) is called a storage function, and if such a storage function exists, then system(16) is called dissipative with respect to the supply rate $\frac{1}{2}(\gamma^2\|\mathbf{w}(t)\|^2 - \|\mathbf{z}(t)\|^2)$. If $V(\mathbf{x})$ is differentiable, we can rewrite (20) as

$$\dot{V}(\mathbf{x}) \le \frac{1}{2}(\gamma^2\|\mathbf{w}(t)\|^2 - \|\mathbf{z}(t)\|^2) \tag{21}$$

which combined with (16) gives

$$\frac{\partial V}{\partial \mathbf{x}}^T(\mathbf{x})(\mathbf{f}(\mathbf{x}) + \mathbf{g}(\mathbf{x})\mathbf{u} + \mathbf{a}(\mathbf{x})\mathbf{w}) - \frac{1}{2}\gamma^2\|\mathbf{w}(t)\|^2 + \frac{1}{2}\|\mathbf{z}(t)\|^2 \le 0 \tag{22}$$

We call the left hand side of (22) the energy Hamiltonian H. We perform a min-max operation on H following the differential game analogy. By solving $\frac{\partial H}{\partial \mathbf{w}} = 0$ and $\frac{\partial H}{\partial \mathbf{u}} = 0$ we obtain the optimum maximizing disturbance $\mathbf{w}^*$ and minimizing input $\mathbf{u}^*$ as:

$$\mathbf{w}^* = \frac{1}{\gamma^2}\mathbf{a}^T(\mathbf{x})\frac{\partial V}{\partial \mathbf{x}},\ \mathbf{u}^* = -\mathbf{g}^T(\mathbf{x})\frac{\partial V}{\partial \mathbf{x}} \tag{23}$$

which provides the saddle point property

$$H(\mathbf{x}, \frac{\partial V}{\partial \mathbf{x}}^T, \mathbf{w}, \mathbf{u}^*) \le H(\mathbf{x}, \frac{\partial V}{\partial \mathbf{x}}^T, \mathbf{w}^*, \mathbf{u}^*) \le H(\mathbf{x}, \frac{\partial V}{\partial \mathbf{x}}^T, \mathbf{w}^*, \mathbf{u}) \tag{24}$$

By substituting (23) in (22), we obtain the Hamilton-Jacobi inequality

$$\frac{\partial V}{\partial \mathbf{x}}(\mathbf{x})\mathbf{f}(\mathbf{x}) + \frac{1}{2}\frac{\partial V}{\partial \mathbf{x}}[\frac{1}{\gamma^2}\mathbf{a}(\mathbf{x})\mathbf{a}^T(\mathbf{x}) - \mathbf{g}(\mathbf{x})\mathbf{g}^T(\mathbf{x})]\frac{\partial V}{\partial \mathbf{x}}^T + \frac{1}{2}\mathbf{h}(\mathbf{x})\mathbf{h}^T(\mathbf{x}) \le 0 \tag{25}$$

If the system is reachable from $\mathbf{x}_0$ then the storage function (19) is finite, and if it is also smooth then it is also a solution of the Hamilton-Jacobi-Isaac equation

$$\frac{\partial V}{\partial \mathbf{x}}(\mathbf{x})\mathbf{f}(\mathbf{x}) + \frac{1}{2}\frac{\partial V}{\partial \mathbf{x}}[\frac{1}{\gamma^2}\mathbf{a}(\mathbf{x})\mathbf{a}^T(\mathbf{x}) - \mathbf{g}(\mathbf{x})\mathbf{g}^T(\mathbf{x})]\frac{\partial V}{\partial \mathbf{x}}^T + \frac{1}{2}\mathbf{h}(\mathbf{x})\mathbf{h}^T(\mathbf{x}) = 0 \tag{26}$$

In the game theoretic formulation, the objective function on which the players perform min-max is

$$J(\mathbf{u},\mathbf{w}) = \int_0^T (\mathbf{z}'\mathbf{z} - \gamma^2 \mathbf{w}'\mathbf{w})dt \tag{27}$$

The solution of the game theoretic formulation is given by (23) in conjunction with (26).

The solution for the standard infinite time horizon optimal control problem can be obtained from this by eliminating the disturbance player w(t). This can be achieved by taking the limit $\gamma \rightarrow \infty$ in the Hamilton-Jacobi inequality. Hence an optimal control problem with a feedback solution for minimizing

$$J(\mathbf{u}) = \int_0^T (\mathbf{z}'\mathbf{z})dt \tag{28}$$

The solution of this is given by

$$\mathbf{u}^* = -\mathbf{g}^T(\mathbf{x})\frac{\partial V}{\partial \mathbf{x}} \tag{29}$$

where V(x) is the solution of the Hamilton Jacobi equation

$$\frac{\partial V}{\partial \mathbf{x}}(\mathbf{x})\mathbf{f}(\mathbf{x}) + \frac{1}{2}\frac{\partial V}{\partial \mathbf{x}}[-\mathbf{g}(\mathbf{x})\mathbf{g}^T(\mathbf{x})]\frac{\partial V}{\partial \mathbf{x}}^T + \frac{1}{2}\mathbf{h}(\mathbf{x})\mathbf{h}^T(\mathbf{x}) = 0 \tag{30}$$

<u>Polynomial Approximation Method for Solving Hamilton-Jacobi equation and inequality</u>:
Set

$$V(\mathbf{X}) = V^{[2]}(\mathbf{X}) + V^{[3]}(\mathbf{X}) \tag{31}$$

where $V^{[2]}(\mathbf{X})$ contains second order terms and $V^{[3]}(\mathbf{X})$ contains third order terms. For solving (30), we can substitute (31) in (30), and solve for similar order terms. The details of using the polynomial approximation technique, which provides local results are shown in references [10] and [11]. A software, Nonlinear Systems Toolbox [12], is available which solves the Hamilton-Jacobi equation using the power series.

<u>Measurement Feedback Control</u>: The above described solutions (23), (29) are valid when the full state is available for feedback, i.e. the full state is measured directly. On the other hand, in many cases, the full state is not available. In those cases, it needs to be found out if the partial measurement available renders the system observable; in other words, can the state variables be estimated from the measured outputs. If the system is observable, then we can design state observers which process the measured outputs and provide best (in some sense) estimates of the state variables, which then can be used in the controllers. There is a good amount of literature on the topic of linear and nonlinear observers. In linear systems, Luenberger observer and Kalman filters have been used effectively [13]. Reference [14] provides a good survey of nonlinear observers.

For dynamic traffic assignment problems using feedback control, normally state variables like traffic flow or traffic density are measured. Other variables like split parameters, origin-destination flows have to be estimated. If information about split factors and origin-destination flows is available through communication with vehicles (such as by using GPS, cellular communication, etc.) then it becomes a full state feedback control problem, but at the present level of applied technology, these variables have to be estimated. There has been some effort at building Kalman filter based observers for origin-destination trip table estimation [14, 15], but the authors have not seen any work in the area of estimating the composition rates. This will be area of further research by the authors, which would then in conjunction with the work presented in this paper provide an immediately deployable DTA scheme. In the meantime, however, this solution is highly attractive for off-line simulation studies also and for preliminary design for deployable systems, which would work with state observers for real time deployable feedback systems.

**DTA Problem using Link Based Model**

The dynamics of the link based model can be written in form (15) by using equations (1), (2), (5-10) and either (11) or (12). In this paper we will deal with equations of type (12) instead of (11) for composition rate dynamics. In the link based model, the link densities can be used directly to formulate the system cost. For instance, if we are trying to minimize the weighted cost of input and the user-equilibrium travel cost, we can write the variable $\mathbf{z}(t)$ as

$$\mathbf{z}(t) = \begin{bmatrix} w\,h(\mathbf{x}) \\ \mathbf{u} \end{bmatrix} \tag{32}$$

where w is the relative weight on the input. We can also minimize the weighted cost of input and the total travel time experienced by all the travelers using (32). If we assume that travel time on a link is given by the quotient of the division of its length with the average velocity on it. On that basis, h(x) will be

$$h(\mathbf{x}) = \sum_{i \in P} e_i, \quad \text{where } e_i = \sum_{k=1}^{k=\ell_{nj}} (\Delta_k - \Delta_{k+1})^2, \tag{33}$$

$$\text{and } \Delta_k = \sum_{i \in r_k} \delta_i \rho_i / Q_i$$

for user-equilibrium, and

$$h(\mathbf{x}) = \sum_{i \in M} \Delta_i, \quad \text{where } \Delta_i = \delta_i \rho_i / Q_i \text{ for system optimal}$$

We have taken $\ell_{nj} + 1$ to be same as 1. The symbol $\Delta_k$ indicates the total travel time on the kth alternate route starting from the node I, P is the set of all the node-destination pairs nj, and $r_k$ is the set of all links in the kth alternate route.

**DTA Problem using Route Based Model**

Route based system dynamics model is obtained by combining (1), (2), (5-8), (13) and (14). In this case also, the system cost is a weighted function of the input and state dependent cost. Since the state variables are different in this case, the state dependent cost will have to be written in a different form. We can take $z(t)$ to be

$$\mathbf{z}(t) = \begin{bmatrix} w\,h(\mathbf{x}) \\ \mathbf{u} \end{bmatrix} \tag{34}$$

where w is the relative weight on the input. For the route based model $h(\mathbf{x})$ can be written as

$$h(\mathbf{x}) = \sum_{i \in P} e_i, \quad \text{where } e_i = \sum_{k=1}^{k=\ell_{nj}} (\Delta_k - \Delta_{k+1})^2, \tag{35}$$

$$\text{and } \Delta_k = \sum_{i \in r_k} \delta_i [\sum_{j \in S_m} \rho_{ij}] / Q_i$$

for user-equilibrium, and

$$h(\mathbf{x}) = \sum_{i \in M} \delta_i [\sum_{j \in S_m} \rho_{ij}] / Q_i \quad \text{for system optimal}$$

**Feedback Control for the Traffic**

When the complete extended system which includes the dynamics of the OD flows as well as those of the split factors, so that the input vector consists of all the rate change of split factors, is written in form (16), then the feedback control for the system is given by (23) combined with solution of the Hamilton-Jacobi inequality, where appropriate substitution of $h(\mathbf{x})$ is made from (32) or (34). Specifically, in (25) wherever $h(\mathbf{x})$ appears, it will be replaced by $w\,h(\mathbf{x})$.

# 4. Sample Problem

We show the concepts developed in this paper using a simple triangular network consisting of three nodes. This problem has all the relevant elements of the network which are important from the modeling perspective. It is a multi-destination network with two alternate routes for each origin destination pair. The network is described in terms of the mathematical variables below.

**System Network**

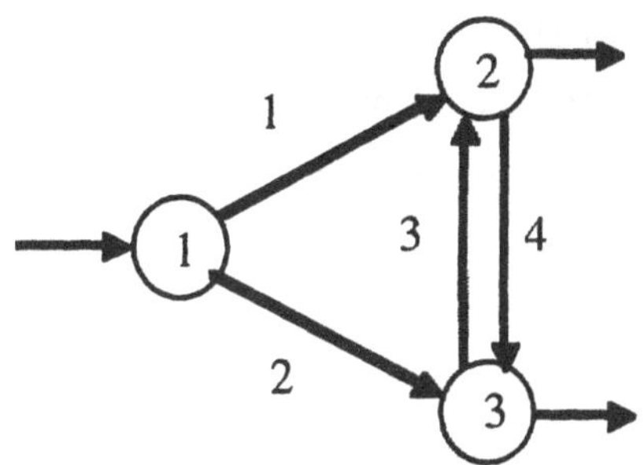


**Figure 8.3**: *Sample Network*

The sample network is shown in the directed graph in figure 3. The network has one origin node, node 1, and two destination nodes, node 2 and node 3. Hence, the origin and destination sets are given by $O=\{1\}, D=\{2,3\}$. For the OD pair $\{1,2\}$, there are two alternate routes and are given by $\ell_{1,2}=2, L^1_{1,2}=\{1\}, L^2_{1,2}=\{2,3\}$. Similarly for OD pair $\{1,3\}$, $\ell_{1,3}=2, L^1_{1,3}=\{1,3\}, L^2_{1,3}=\{2\}$.

There are two independent destination based splitting rates for the sample problem as shown in Table 8.1, but only one independent node based splitting rate $\beta^1_1$.

**Table 8.1** *Destination Based Splitting Rates for Sample Problem*

| From Node | To Destination Node | Splitting Rates | Independent Splitting Rates | No. |
|---|---|---|---|---|
| 1 | 2 | $\beta^1_{12}, \beta^2_{12}$ | $\beta^1_{12}$ | 1 |
| 1 | 3 | $\beta^1_{13}, \beta^2_{13}$ | $\beta^1_{13}$ | 2 |

**Table 8.2** *Composition Rates for Sample Problem*

| Link | Composition Rates | Independent Composition Rates | No. |
|---|---|---|---|
| 1 | $\gamma_{12}, \gamma_{13}$ | $\gamma_{12}$ | 1 |
| 2 | $\gamma_{22}, \gamma_{23}$ | $\gamma_{22}$ | 2 |

The destination based flows entering the origin node 1 are given by $q_{12}=d_{12}, q_{13}=d_{13}$. The entrance flows on the four links of the network are related to various split factors, exiting composition rates, exit flows, and OD flows as shown below in (36).

$$q_1 = \beta^1_{12} d_{12} + \beta^1_{13} d_{13}, q_2 = (1-\beta^1_{12}) d_{12} + (1-\beta^1_{13}) d_{13} \qquad (36)$$
$$q_3 = Q_2 \Gamma_{22}, q_4 = Q_1 \Gamma_{13}$$

The entering composition splits in various links of the network are shown below.

$$\gamma_{12} = \beta^1_{12} \frac{q_{12}}{q_1}, \gamma_{13} = 1 - \gamma_{12}, \gamma_{22} = \beta^2_{12} \frac{q_{12}}{q_2}, \gamma_{23} = 1 - \gamma_{22}, \gamma_{32} = \qquad (37)$$

**System Dynamics**

As mentioned is section 2, we can model the sample network either by using link based variables or route based variables. We can also model the system using destination based splits or node based splits. For the sake of brevity of this paper, we will consider only the destination based split case for the link based model.

For the sample network, there are four links, and hence there are four state variables which are the traffic densities on these four links. Since, there are traffic splits only at node 1, we see the introduction of split inputs in the differential equations for link 1 and link 2. We also see the same equations having the OD flows.

$$\dot{\rho}_1(t) = \frac{1}{\delta_1} [\beta^1_{12} d_{12} + \beta^1_{13} d_{13} - q_{max,1} [1 - e^{-\rho_1(t)/\kappa_1}]] \qquad (38)$$

$$\dot{\rho}_2(t) = \frac{1}{\delta_2} [(1-\beta^1_{12}) d_{12} + (1-\beta^1_{13}) d_{13} - q_{max,2} [1 - e^{-\rho_2(t)/\kappa_2}]]$$

$$\dot{\rho}_3(t) = \frac{1}{\delta_3} [q_{max,2} [1 - e^{-\rho_2(t)/\kappa_2}] \Gamma_{22} - q_{max,3} [1 - e^{-\rho_3(t)/\kappa_3}]]$$

$$\dot{\rho}_4(t) = \frac{1}{\delta_4} [q_{max,1} [1 - e^{-\rho_1(t)/\kappa_1}] \Gamma_{13} - q_{max,4} [1 - e^{-\rho_4(t)/\kappa_4}]]$$

The composition rate dynamics can be modeled using either delay model or filter model structure. For this sample problem, we are using the filter dynamic relationship, which is shown below.

$$\dot{\Gamma}_{12}(t)=\alpha_1[\frac{\beta^1_{12}d_{12}}{\beta^1_{12}d_{12}+\beta^1_{13}d_{13}}-\Gamma_{12}]+\Gamma_{12} \tag{39}$$

$$\dot{\Gamma}_{13}(t)=\alpha_1[1-\Gamma_{13}-\frac{\beta^1_{12}d_{12}}{\beta^1_{12}d_{12}+\beta^1_{13}d_{13}}]+\Gamma_{13}$$

$$\dot{\Gamma}_{22}(t)=\alpha_2[\frac{\beta^2_{12}d_{12}}{(1-\beta^1_{12})d_{12}+(1-\beta^1_{13})d_{13}}-\Gamma_{22}]+\Gamma_{22j}$$

$$\dot{\Gamma}_{23}(t)=\alpha_2[1-\Gamma_{23}-\frac{\beta^2_{12}d_{12}}{(1-\beta^1_{12})d_{12}+(1-\beta^1_{13})d_{13}}]+\Gamma_{23}$$

Now, we will need to extend the state variable vector to introduce the OD dynamics and to make the rate of change of split factors as the input, so that in the optimal control formulation, we can make that change bounded, as discussed in section 2. Following Okutani's model [15] and taking diagonal elements in the transition matrix for OD dynamics, we can decouple the OD dynamics as

$$\dot{d}_{12}(t)=\alpha_1 d_{12}(t),\ \dot{d}_{13}(t)=\alpha_2 d_{13}(t) \tag{40}$$

where $\alpha_1$ and $\alpha_2$ are time varying parameters. Now using parameters in the system model which are dependent explicitly on time, renders the system time varying. In order to obtain a stationary solution of the Hamilton Jacobi equation, we require the system to be time invariant. This can be obtained by modeling the OD dynamics in greater detail. If we look at the plot of OD flows in a network, we observe periodicity with usually multiple peaks every day. This periodic nature of the flows can be modeled by higher order time invariant dynamics. For instance, a periodic sine wave is a solution of a second order linear time invariant system with purely imaginary eigenvalues. We can represent a nominal OD flow using Fourier series, which is a sum of sinusoidal functions representing the fluctuations of the system, and then obtain a time invariant model of the flow. This approach of representing the OD flows by time invariant higher order dynamics is one approach we can use in DTA when the assignment has to be carried out for a long period of time which allows the periodic nature of the OD flows to appear. On the other hand, if the assignment needs to be carried out for a shorter period of time, say a couple of hours, then we can use the Kalman filter to find out the value of the OD flows, and then employ that feedback control law which is designed for that OD flow value. This scheme is called gain scheduling. This scheme divides the range of OD flows into a finite number of cells each with an average flow value, and for each value we can design a feedback control off-line. In real time, when a value of the OD flow is identified, then the controller designed for that value is used till the value of the flow moves out of that cell. In the gain scheduling case, the flow values are considered constant and equal to the average value of that cell till the actual OD flow value remains in that cell.

We now introduce new input variables as

$$\dot{\beta}^1_{12}=u_1,\ \dot{\beta}^1_{13}=u_2 \tag{41}$$

Now, we can combine the dynamics (38, 39, 41) and rewrite the combined equations in a matrix form, which shows the system in an input-affine form using the new input variables. The integrated model is shown in (42). Here we are showing the design steps of one of the gain scheduled controllers.

$$
\begin{bmatrix} \dot\rho_1(t) \\ \dot\rho_2(t) \\ \dot\rho_3(t) \\ \dot\rho_4(t) \\ \dot\Gamma_{12}(t) \\ \dot\Gamma_{13}(t) \\ \dot\Gamma_{22}(t) \\ \dot\Gamma_{23}(t) \\ \dot\beta^1_{12} \\ \dot\beta^1_{13} \end{bmatrix} =
\tag{42}
$$

$$
\begin{bmatrix}
\frac{1}{\delta_1}[-q_{max,1}[1-e^{-\rho_1(t)/\kappa_1}]]+\frac{d_{12}}{\delta_1}\beta^1_{12}+\frac{d_{13}}{\delta_1}\beta^1_{13} \\
\frac{1}{\delta_2}[d_{12}+d_{13}-q_{max,2}[1-e^{-\rho_2(t)/\kappa_2}]]-\frac{d_{12}}{\delta_1}\beta^1_{12}-\frac{d_{13}}{\delta_1}\beta^1_{13} \\
\frac{1}{\delta_3}[q_{max,2}[1-e^{-\rho_2(t)/\kappa_2}]\Gamma_{22}-q_{max,3}[1-e^{-\rho_3(t)/\kappa_3}]] \\
\frac{1}{\delta_4}[q_{max,1}[1-e^{-\rho_1(t)/\kappa_1}]\Gamma_{13}-q_{max,4}[1-e^{-\rho_4(t)/\kappa_4}]] \\
\alpha_1[\frac{\beta^1_{12}d_{12}}{\beta^1_{12}d_{12}+\beta^1_{13}d_{13}}-\Gamma_{12}]+\Gamma_{12} \\
\alpha_1[1-\Gamma_{13}-\frac{\beta^1_{12}d_{12}}{\beta^1_{12}d_{12}+\beta^1_{13}d_{13}}]+\Gamma_{13} \\
\alpha_2[\frac{\beta^2_{12}d_{12}}{(1-\beta^1_{12})d_{12}+(1-\beta^1_{13})d_{13}}-\Gamma_{22}]+\Gamma_{22j} \\
\alpha_2[1-\Gamma_{23}-\frac{\beta^2_{12}d_{12}}{(1-\beta^1_{12})d_{12}+(1-\beta^1_{13})d_{13}}]+\Gamma_{23} \\
0 \\
0
\end{bmatrix} +
$$

$$
\begin{bmatrix} 0 & 0 \\ 0 & 0 \\ 0 & 0 \\ 0 & 0 \\ 0 & 0 \\ 0 & 0 \\ 0 & 0 \\ 0 & 0 \\ 1 & 0 \\ 0 & 1 \end{bmatrix}
\begin{bmatrix} u_1 \\ u_2 \end{bmatrix}
$$

which is in the form

$$\dot{\mathbf{x}} = \mathbf{f}(\mathbf{x}) + \mathbf{g}(\mathbf{x})\mathbf{u} \tag{43}$$

where $\mathbf{u} = [u_1 \quad u_2]^T$, $\mathbf{f}(\mathbf{x})$, $\mathbf{g}(\mathbf{x})$ can be obtained from (42) and the state variable is given by

$$\mathbf{x} = [\rho_1(t) \quad \rho_2(t) \quad \rho_3(t) \quad \rho_4(t) \quad \Gamma_{12}(t) \quad \Gamma_{13}(t) \quad \Gamma_{22}(t)$$
$$\Gamma_{23}(t) \quad \beta_{12}^1 \quad \beta_{13}^1]^T$$

Now, we can address explicitly the uncertainties of the system (43) and write the complete model including uncertainties as (16). For instance, we could consider the uncertainties in the parameters of (42) and then take these out and equate that to **a(x)w** term of (16), after dividing the uncertainty into **a(x)** and **w**. On the other hand, if the nominal model is accurate, then **w=0**, which implies that one of the two players of the nonlinear $H_\infty$ game is not present. The controller design for both of these cases is given in the next section.

Now, in order to formulate the problem, we choose the variable z(t) from (32), which combined with (33) for this problem gives

$$\mathbf{z}(t) = \begin{bmatrix} \sum_{nj=(1,2),(1,3)} \sum_{k=1}^{k=\ell_{nj}} \left(\sum_{i \in r_k} \delta_i \rho_i / Q_i - \sum_{i \in r_{k+1}} \delta_i \rho_i / Q_i\right)^2 \\ \mathbf{u} \end{bmatrix} \text{ for user equilibrium} \tag{44}$$

$$\mathbf{z}(t) = \begin{bmatrix} w \sum_{i=1}^{4} \delta_i \rho_i / q_{max,i} [1 - e^{-\rho_i / \kappa_i}] \\ \mathbf{u} \end{bmatrix} \text{ for system optimal}$$

Note that there are two alternate routes each for the two node-destination (1,2) and (1,3) pairs. Figure 8.4 shows the two alternate routes for the (1,2) pair.

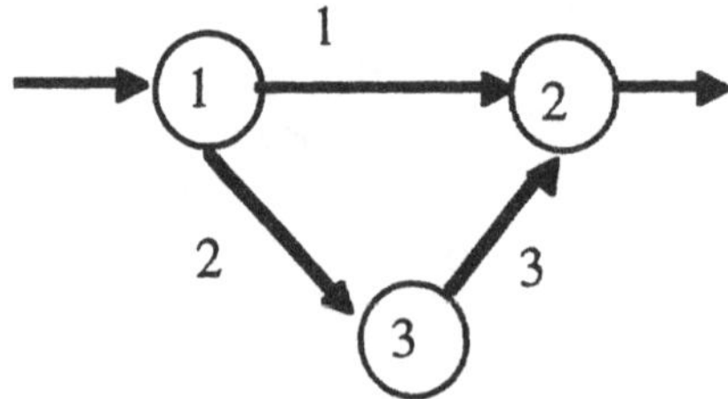


**Figure 8.4**: *Alternate Routes for the Node destination pair (1,2)*

**Feedback Control Design**

The solution of the feedback control is given by $\mathbf{u}^* = -\mathbf{g}^T(\mathbf{x})\frac{\partial V}{\partial \mathbf{x}}$, where V is the solution of the Hamilton Jacobi inequality (25). For this problem, making substitutions from (42) here, gives the control input as

$$\mathbf{u}^* = \begin{bmatrix} -\dfrac{\partial V}{\partial \beta_{12}^1} \\ -\dfrac{\partial V}{\partial \beta_{13}^1} \end{bmatrix} \tag{45}$$

which gives the solution to this problem as two differential equations

$$\dot{\beta}_{12}^1 + \frac{\partial V}{\partial \beta_{12}^1} = 0, \quad \text{and} \quad \dot{\beta}_{13}^1 + \frac{\partial V}{\partial \beta_{13}^1} = 0 \tag{46}$$

The scalar V is obtained from the solution of (25) that can be solved using the power series method.

# 5. Summary

In this paper we present a real-time on-line feedback control solution for the network wide dynamic traffic assignment problem using user equilibrium. The solutions, which are based on nonlinear $H_\infty$ design, are shown for link based as well route based models, and it is also shown how the problem can be modeled and solved using destination based and route based split factors. The control design for a sample network is also presented.

# 6. Exercises

## Questions

Question 1:How is a network level problem different than point diversion problem?
Question 2:How can you represent a point diversion problem using the network level modeling?
Question 3: In the control algorithms developed in this chapter which calculations are to be performed off-line and which ones are to be performed on-line?
Question 4: What equation has to be solved in order to obtain the controller using the design presented in this chapter?
Question 5: How can the equation of Question 4 be solved?

## Problems

*Problem 1*

Write down the system dynamics in terms of link dynamics and node dynamics, for the problem shown in Figure 8.5.

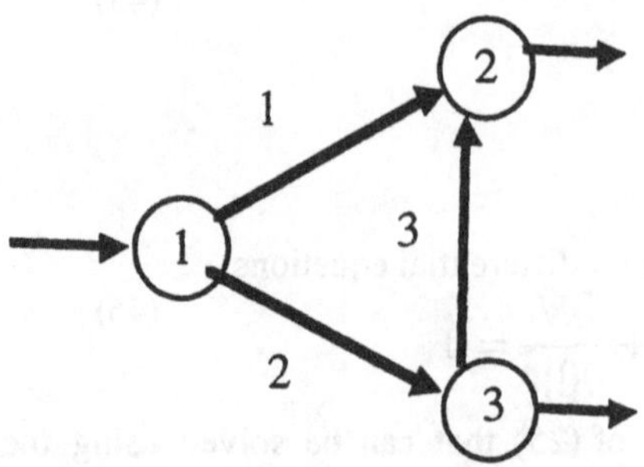


**Figure 8.5**: *Problem 1 Network*

*Problem 2*

Design user-equilibrium and system optimal controller for the problem shown in Figure 8.5.

# 7. References


(1) J. A. Ball, J. W. Helton, and M. L. Walker, $H_\infty$ Control for Nonlinear Systems with Output Feedback, IEEE Trans. on Automatic Control, Vol. 00, No. 00, April 1993.

(2) W. M. Lu, and J. C. Doyle, $H_\infty$ Control of Nonlinear Systems via Output Feedback: Controller Parametrization, IEEE Trans. on Automatic Control, Vol. 39, No. 12, December 1994.

(3) Isidori, A., and Kang, W., "$H_\infty$ Control via Measurement Feedback for General Nonlinear Systems", IEEE TRANS. on Aut. Control, vol. 40, No. 3, March 1995.

(4) van der Schaft, A. J., "Nonlinear State Space $H_\infty$ Control Theory", Perspectives in Control, Feb. 1993.

(5) J. A. Ball, and J. W. Helton, $H_\infty$ Control for Nonlinear Plants: Connections with Differential Games, in Proc. 28th Conf. Decision and Control, Tampa, FL, Dec. 1989, pp. 956-962.

(6) J. C. Willems, Dissipative Dynamical Systems, Part I: General Theory, Arch. Rat. Mech. Anal., vol. 45, pp. 321-351, 1972.

(7) T. Basar and G. J. Olsder, "Dynamic Noncooperative Game Theory", New York: Academic, 1982.

(8) A Friedman, "Differential Games", Wiley-Interscience, 1971.

(9) Papageorgiou, M., 'Dynamic Modeling, Assignment and Route Guidance Traffic Networks', Transportation Research-B, Vol. 24B, No. 6, 471-495, 1990.

(10) Brekht, E. G. Al', "On the Optimal Stabilization of Nonlinear Systems", PMM-J. Appl. Math. Mech., vol. 25, no. 5, pp. 836-844, 1961.

(11) Krener, A. J., "Optimal Model Matching Controllers for Linear and Nonlinear Systems", IFAC Nonlinear Control Systems Design, Bordeaux, France, 1992.

(12) Krener, A. J., "The Nonlinear Systems Toolbox", available from SCAD, http://scad.utdallas.edu/scad/software.html
(13) Stengel, Robert F., "Optimal Control and Estimation" Dover Publications, Inc., New York, 1986.
(14) Misawa, E. A., and Hedrick, J. K., "Nonlinear Observers: A State of the Art", ASME Journal of Dynamic Systems Measurement and Control, Sept. 1989.
(15) Okutani, I., "The Kalman Filtering Approach in Some Transportation and Traffic Problems", Intl. Symposium on Transportation and Traffic Theory, N. H. Gartner and N. H. M. Wilson (eds.), Elveiser Science Publishing Co. Inc., 397-416, 1987.
(16) Ashok, Ben-Akiva, M. E., "Dynamic Origin-Destination Matrix Estimation and Prediction for Real-Time Traffic Management Systems", Transportation and Traffic Theory, C. F. Darganzo (ed.), Elveiser Science Publishing Co., Inc., 465-484, 1993.

# INDEX

## E



## F



## G



## H



## I



## K

## L



## M



## N



## O



## P

## Q



## R



## S



## T



## U

Zeitfracht Medien GmbH
Ferdinand-Jühlke-Straße 7
99095 Erfurt, Deutschland
produktsicherheit@kolibri360.de